ARNOLD MARSDEN

Safety First! Really?

How to be a Credible Safety Leader

Contents

Preface

"Safety First!"

Really? Are you sure?

"How dare you ask? That's blasphemy! Of course, Safety is First. How could it be anything but First? In fact, our goal is Zero Harm; everyone goes home safe. Ask anyone in the organization, and they'll tell you the same thing."

Will they? Have you asked them how they really feel? Not just your direct reports or the safety team, but the people holding the tools, sweating in the scorching sun, trying to meet all your other high expectations?

You repeat it often. It is plastered all over the office and at the worksites. But do you really mean it? Look at yourself in the mirror. Specifically, think about your actions and decisions over the last year. Do they all make the workplace safer? How many improve safety versus improving revenue, cost, or product quality? Did you ever sacrifice safety for production? Even a little bit? If you bear with me, this book will be your mirror. Then you'll really know whether Safety is First. And if not, it will help you demonstrate your safety commitment to your organization, whatever your vision is.

But first, let me tell you a story that inspired this book.

My daughter was studying to be an engineer and working on a summer internship for a chemical company. At the time, I had just passed my thirty-year anniversary working for a large oil and gas company, mostly as a Health, Safety, and Environment professional. During one of her weekend visits home, she, my wife, and I were comparing and contrasting safety programs in the two companies. My wife mentioned a video she saw on YouTube by Mike Rowe claiming Safety was Third, not First. You've probably heard of Mike Rowe; he was the host of the television show, *Dirty Jobs*, among other things. Hmmm…he works with people in danger all the time. It might be worth a look.

I was on the agenda for the safety moment at our plant leadership team meeting the following week. Our leadership team used the admirable practice of rotating the responsibility for the safety moment versus always looking to the safety person when no volunteers stepped up.

Maybe a discussion of this video will do?

The video was just over three minutes, concise enough for a Safety Moment. And it was provocative–sure to spur more discussion than flipping through a few slides on *Summer Safety Tips*. Despite the title, Mike Rowe was not downplaying the importance of safety. In fact, he was doing the opposite. My takeaways from the video were: (1) businesses and leaders have multiple priorities, (2) overusing slogans such as Safety First can lead workers to assume the company will take care of them, and 3) your safety is up to YOU! He quoted a crab boat captain from his work on the *Deadliest Catch*, "I'm not here to get you home alive, I'm here to get you home rich!" By being provocative and claiming safety was third, perhaps he would shake up those who had become complacent.

This should be a great safety moment, right? Full of spirited discussion versus nodding heads and impatient glares.

Not! At least not for that team on that day. It was spirited alright, but not so honest.

Once the team got past the initial shock of the blasphemy, I thought the video would lead to an insightful discussion about the dilemmas we face in balancing all our value drivers. You know, value drivers like safety, cost, schedule, profit, customer satisfaction, and quality. To add to the stakes, Monday's meeting would be the first for our new plant manager. Our paths had crossed a couple times in the past, so it wouldn't be his first impression of me, but important nonetheless.

On Monday morning, I arrived in the conference room ten minutes early to make sure the video and audio were working. Just after I clicked *Play*, the operations manager walked in and looked at the video screen before saying, "Hello." I stopped the video after thirty seconds.

The operations manager raised his eyebrows. "This should be interesting."

Uh Oh! Too late to back out now.

Once everyone had gathered, the operations manager, who was chairing the day's meeting, said with a smirk, "Arnold has an interesting safety moment for us today."

I sensed that he expected fireworks. Everyone was silent during the short video, but I noticed the squinting eyes, shaking heads, and stares directed toward the tabletop. The silence continued after the video ended. Not the typical silence after a safety moment that indicates, "OK, let's move on to the next agenda item." No, this silence and the mannerisms around the table seemed to indicate a mix of disbelief and disgust. How could the safety guy play such a thing?

After fifteen seconds of slumping further and further in my chair, the new plant manager said, "We have an obligation to take care of our people. We can't leave it all to them. Their safety is our responsibility."

"How could you ever justify that safety is third?" asked the engineering manager.

The operations manager had been the first to see the video, but he was the third to comment. "I definitely wouldn't want to share this with the frontline."

I tried to dig myself out of my hole and asked, "Do you think all the frontline workers truly believe safety is first? How do you know? Are we sending any signals that might contradict the message?"

My comment generated another flurry of responses.

"It's the first line in our HSE Policy."

"Our target recordable injury rate is zero."

"Everyone knows they are supposed to stop work if it's not safe, and we always support them when they do."

"I don't get the point of this discussion!"

I was only digging a deeper hole for my casket, so I cut my losses and resolved to make my second and third impressions on the new plant manager really good–which I ultimately did.

Well, guess what, the frontline does see it! If they don't see Mike Rowe's video, they can just look around themselves at work every day. They see and hear the mixed messaging he referred to. They nod their head in violent agreement when they hear the words of the crab boat captain. I was in middle management. I saw many actions and decisions inconsistent with the words "Safety First!" Some were even my own. Granted, I was a safety guy. I was looking for them, but they aren't hard to find. My peers saw them. The frontline saw them.

That was the moment of inspiration for me to write this book. What seemed so obvious to me as a safety professional was not so apparent to some of the senior line leaders I worked with, or at least they were not willing to discuss it openly. Most of them were good safety leaders, some even excellent, but their words, actions, and behaviors didn't always show it. If they couldn't discuss the issue with their peers, then how would they handle similar challenges by frontline leaders? I saw examples every week. They were generally black and white, a good thing since I am color blind. I assumed it was obvious to the leaders I worked with as well. And if they didn't notice it, how could the leaders above them see it when they were even further removed from the frontline?

So what? What's the harm in a few mixed messages?

Mixed messages lead to a loss of a leader's credibility, which leads to mixed performance in the organization. Some recognize the hazards, others don't. Some implement hazard controls, others don't. And you know what follows…injuries. In other words, lives are at stake!

Perhaps what I observed was not so obvious, and others may learn from my insights. Perhaps they can benefit from me holding up a mirror to their face, so they could take a closer look.

Are you humble enough for the challenge? By the way, humility is one of the traits of an excellent safety leader!

1

Introduction

How do you describe yourself as a safety leader? Strong, weak, or good enough? Consistent or inconsistent? Proactive or reactive? Caring or business-like?

How do people in your organization describe your safety leadership? Do you know? Do you REALLY know?

Are they the same?

After all, the people in your organization decide what kind of safety leader you are, not you. Sure, you can decide what kind of safety leader you *want to be*, but the people in your organization decide what type of safety leader *you are* by interpreting your words, actions, decisions, and behaviors.

Don't fool yourself into believing you are only judged as a safety leader while giving the safety update at the beginning of a town hall meeting, going on a safety walk in the facility, and handing out coffee mugs for reaching an injury-free milestone. People are listening to every word and watching every move. One impulsive action or comment inconsistent with your professed safety vision can undo a dozen orchestrated positive messages. Those day to day sound bites, reactions, decisions, and even body language are the cues that expose your true feelings and motivations. That's why safety leadership is so hard. It's a 24/7 job; you are always on stage. And the stakes are as high as they can get, people's lives. For most leaders, safety is the hardest part of their job. If it doesn't feel that way, perhaps you should do more.

That being said, few people expect their leader to be perfect–just like the leader shouldn't expect the mechanics or operators on the frontline to be perfect. Most are not expecting a risk-free workplace. But they do expect their leaders to provide a safe work environment. They expect them to articulate what their safety vision is and to act accordingly. Otherwise, a leader will foster skepticism or distrust, which leads to complacency, inconsistent work practices, and even defiance. And ultimately, injuries.

The purpose of this book is to show leaders how their words, actions, decisions, and behaviors determine their credibility and effectiveness as safety leaders.

In this book, I will:

- Show the importance of a leader understanding their safety vision in detail and fully committing to it.
- Describe the key traits of an excellent safety leader and the actions they can take to demonstrate them.
- Share specific examples of leaders supporting and contradicting their safety vision and how they impact safety behaviors in the organization.
- Present options for a leader to gather feedback on their actual safety messaging.
- Describe how leaders at all levels of an organization can work together to drive continuous improvement in safety.

The purpose of this book is NOT to tell you what your safety vision should be, what Zero Harm really means, and whether it can be achieved. However, I will cover some of the options and the obstacles they can create in becoming a credible safety leader.

The target audiences for this book are leaders who are already committed to safety excellence and seeking to improve and the HSE professionals who help them. If you are not committed, and safety really is third, then this book may not be for you. Those leaders may read this book so they can get better at faking it, but from my experience, no one can fake it all the time, or even 80% of the time. There are endless opportunities to get it wrong and too many eyes watching and ears listening.

This book is based on experience, not scientific theory and research. You won't find a long list of references at the end. Other authors have done a much better job of presenting the theory. However, some of those books lack the practical implications that only decades of observations and experience bring. I spent over thirty-five years as an HSE professional observing the safety (and environmental) leadership behaviors of leaders at all levels in many companies with high hazard work. I've seen their impacts on the rest of the organization. I coached and was coached. And I learned the hard way, through my own behaviors.

My objective is to expedite your journey to being an excellent safety leader, thereby reducing the risk of injury to people in your organization along the way.

The Talk

I'm sure you've heard the phrase, *Walk the Talk*. It's one of my favorites. I am a firm believer in leading by example. But let's add a couple of complimentary phrases to make the thought complete:

- Know the Talk
- Walk the Talk
- Hear the Talk

Before you can walk the talk, you need to know what the talk is. In other words, what is your safety vision? What are you trying to accomplish? Is it zero Injuries? First Aids or Fatalities? Is it no serious injuries? What is serious? Is it the presence of effective safety controls? Is it safety above all else or safety on par with another aspect of business delivery?

Tough questions, but important ones that you need to know the answers to.

Only when you know what you want the organization to accomplish can you walk–or act–accordingly. As you will see, walking the talk is hard enough when your vision and goals are clear. If not, it's impossible. This is *Knowing the Talk*. I am not here to tell you what your vision should be. That is a very personal decision, and one that should result from collaboration with your leadership team.

Now you can *Walk the Talk*. People are looking for evidence of their leader supporting their safety vision…or not. Some encounters may not have any signals related to safety, but many times there are, more than most leaders think.

- Did you select the scaffolding company with the best safety performance?
- How were safety leadership behaviors considered when selecting the new operations manager?
- Is safety just a *moment* at the beginning of a leadership meeting or is it integrated into all discussions and decisions where it is relevant?
- Are people rewarded for getting the job done even though they took shortcuts that could have led to an injury?
- Do you assume no news is good news on the safety front? Do you seek others' views on how to improve, no matter how good the recent safety performance?
- How did you react to the latest incident?

We will explore situations like these in detail throughout the rest of the book, mostly through specific examples based on real-world experiences. Employees tend to scrutinize leaders most in two scenarios: when someone is injured and when safety clashes with another business driver, such as cost or production. A leader must speak and act in accordance with their safety vision in these two critical situations and countless others.

Finally, do you seek feedback from others on how aligned your words and actions are with your safety vision? That is *Hearing the Talk*. If a leader *Knows the Talk* and genuinely attempts to *Walk the Talk*, they are already committed to the vision. But the individuals in the organization still need to be convinced by supporting behaviors. They will determine how well you are walking the talk. How do you know what messages they are receiving?

It's like editing a book. By the time I finished the first draft of this book and a few editing passes, I was convinced the content was right and the narrative was clear. I was too close to see the gaps in information, missing commas, repetition, and unclear passages. Editors and beta readers have the distance and independence to provide the needed feedback.

Who can provide you with such feedback on how well you walk the talk? Do you ask open-ended, non-leading questions? Do you actively listen to what they have to say? How do you react when you are surprised by the feedback? How often do you say, "But that's not what I meant!" Well, the message received is more important than the message intended. Are you humble enough to acknowledge the feedback, ask for clarification, and improve in the future? I will cover this in more detail in Chapter 8.

Know the Talk, Walk the Talk, Hear the Talk.

Simple on paper

Safety leadership, and leadership in general, for that matter, is not only difficult to practice but also to describe. Different people in the organization are looking for different traits in a safety leader. Different combinations of traits work for different leaders because of their natural strengths, weaknesses, and passions. But after thirty-five years of observing, scrutinizing, and coaching safety (and HSE) leadership behaviors, two absolutely necessary traits stand out.

SINCERITY plus CONSISTENCY equals CREDIBILITY!

Or said another way, consistent sincerity builds credibility.

Many other specific traits and behaviors support these two key traits. I will cover those in subsequent chapters.

SINCERITY is defined as "honest of mind; freedom from hypocrisy" (Merriam-Webster online dictionary, 2022). In the context of safety, sincerity boils down to how much a leader cares for every individual in the organization and how they demonstrate such care. Are they trying to improve safety performance so everyone goes home safely? Or are they trying to meet a company target or accelerate their move up the corporate ladder? People can tell. As you'll see by example, it's not hard to figure out.

CONSISTENCY is defined as "the quality or fact of staying the same at different times" or "the quality or fact of having parts that agree with each other" (Merriam Webster online dictionary, 2022). Both are relevant in the context of safety leadership. Are a leader's words, actions, and decisions aligned with the safety vision from situation to situation and month to month? Or do they diverge when the going gets tough, i.e., when business sacrifices are needed to maintain their commitment to safety? People know when it is easy to be a good safety leader (the right words at a town hall meeting) versus difficult (shutting down production to reduce a hazard). Most employees use a leader's actions during difficult situations to judge their commitment. It's too easy to fake orchestrated engagements. A leader's true colors show when forced to choose between safety and another key business result.

But for most organizations, it's not enough for one leader to act consistently with their words. All the leaders in an organization must be consistent with each other. Not identical, but consistent. If not, which boss does a person believe and follow?

CREDIBILITY is earned, not taken or given. Merriam-Webster (online dictionary, 2022) defines it as "the quality or power of inspiring belief." In this context, it is earned by consistently demonstrating a sincere commitment to send employees home unharmed every day. Credibility enables a leader to drive positive change and deliver excellent safety performance. But credibility is fragile and easily lost. One false move during a dilemma can replace credibility with skepticism or distrust.

With credibility, workers are more open to changes intended to protect their safety. They will give their leader the benefit of the doubt, even if they don't initially see the connection of a proposed action with the safety vision. They will be inspired to find their own way toward the safety vision. Hence, the *power* included in the definition. Credibility is essential for the continuous improvement needed for excellent safety performance.

OK, this model sounds simple, but we all know it isn't in practice. We will explore this in great detail in Chapters 4-6.

Night and day

The challenging four-year project was nearly complete. The massive, floating offshore oil and gas production platform was in place and secured to the sea floor. The oil and gas processing equipment were onboard, the basic utilities were operational, and the cramped living quarters were functional. But the project team still had months of grueling work to do, connecting all the modular equipment, commissioning the processing equipment, and beginning the start-up sequence. Months of hazardous and complex work with a limited number of people allowed on the platform.

To help the offshore leadership team plan for the safe and efficient execution of this work, a team of engineers, managers, construction specialists, and HSE staff was available onshore 24/7 to assist. Every morning, the key leaders onshore and offshore gathered for a morning call, or planning meeting, to review what had been accomplished overnight, what was planned for the day, and any help the onshore team could provide. Time is money, so the calls were brisk, and they always started with safety.

However, the calls this week were different. The offshore leadership team sensed it from 150 miles away. The previous Construction Manager, Jeff, was based onshore. He had been severely injured in a car accident the previous week and was recovering in the hospital. His deputy, Kate, had stepped into the fire. She had worked on the project for years, but now her role was to make the decisions, not just to offer advice or develop follow-up plans.

The previous morning calls transpired something like this…

"How are we doing on safety?" asked Jeff.

Reggie, the offshore project lead, was ready as always. "No one was hurt yesterday. We had twenty-three safety observations, fourteen were positive, and nine were coaching opportunities. Of the nine, six were related to hand protection, so we asked the crews to review proper glove selection in their toolbox talk this morning.

"Great," said Jeff. "Were you able to install the pipe spool on vessel V-302 last night?"

"No. We need a scaffold to do that, and we didn't have enough scaffolders on board to build it yesterday."

"Are more scaffolders on the way out?" asked Jeff.

"Yes. They're on the first helicopter this morning," the logistics planner chipped in.

"Who's coming back to shore to make room for them? How will we get their work done? We are falling further behind schedule every day…"

And on it went, as they discussed the other work activities on the critical path to start-up. Safety never re-entered the discussion.

But this week…

Reggie was on his two weeks off as part of his 14 days on / 14 days off rotation. His counterpart, Louis, kicked things off. "We had another safe day yesterday, but a scaffolder visited the medic when he felt light-headed. The medic said he was dehydrated. He was much better after getting some fluids and electrolytes and resting in the cool living quarters. We have a problem with the control valves on the fuel gas compressor. We..."

"Hold on, Louis. We'll get back to those valves in a few minutes. What is the scaffolder's name?" Kate asked.

"Oh...Chet...I think."

"How is Chet feeling this morning?"

After an awkward silence, Louis said softly, "I'm not sure, but he must be OK because his supervisor hasn't told me otherwise."

"Please check for me. It's supposed to get even hotter today, so please give the workers an extra break this afternoon."

"Will do," said Louis. "Now back to those valves..."

"One more thing on safety. Yesterday, you mentioned a section of loose grating and said you barricaded the area until it could be repaired. Has it been fixed?"

"Yes. We finished it yesterday afternoon." Louis paused, wondering if he could move on to those troublesome valves. "Now, back to those valves. The technician has been out here for three weeks, and his daughter is getting married on Saturday. He wants to go home tomorrow (Wednesday), in case there are any travel delays. Unfortunately, the supplier can't send out a replacement until Saturday. I think we can fix this tomorrow if he stays another day."

"I understand your concern, but send him home. He can't miss that wedding. The forecast shows storms moving in later in the week, which could delay helicopter flights. Plus, he must be exhausted after three weeks out there. He needs time off anyway."

The Commissioning Lead chimed in. "But Kate, the repair of those valves is on critical path. Every day we miss will delay first production by a day."

"I understand, Sally. But we'll only get there with our people at their best, both mentally and physically. See if the supplier can rearrange their priorities and send a replacement sooner. But make sure the replacement has the required safety training. Let me know if I can help by calling the owner of the company. OK, what's next…"

And on it went, morning after morning.

Just a few well-placed comments and questions by Kate changed the entire complexion of the calls. Jeff may have sincerely cared about the well-being of the crew as much as Kate, but it wasn't very apparent. Jeff rarely asked questions about safety, so Reggie and Louis gave their token safety update and moved on to the work. The first few mornings after Jeff's accident, Kate surprised the leaders offshore with her probing questions on safety. They quickly learned to anticipate her questions and only moved on to work planning and logistics after she was satisfied everyone could finish the day safely. In addition, safety was no longer just a moment, but integrated into all the other discussion where it made sense, like the dilemma involving the valve technician.

Many were surprised when Kate didn't see the valve technician's request as a dilemma. She didn't hesitate. Her reaction demonstrated her commitment to the safety of (and care for) every individual on the platform. More than all the safety signs, Goal Zero jackets, and emails combined. A moment of hesitation would have sent a different message. How do you think her decision made the technician feel? How about the rest of the crew? They probably worked a little harder and smarter that day…and the next. How do you think Jeff would have responded?

Kate consistently showed that she sincerely cared, not just for their safety, but for them as people. She quickly established credibility, which would grow with similar questions and decisions on future calls and visits. The offshore team would be more inclined to have deeper safety discussions of their own and make the right decisions when safety conflicted with schedule or cost. Over the long term–probably even the short term–they would be safer.

Open your mind

This book is filled with stories based on real experiences working with dozens of companies around the world. The facts have been changed, so they are now fictional. Many have been embellished with additional details, but the themes remain the same. The lessons in some are obvious, more subtle in others. I was fortunate to work for and with many organizations with good safety cultures, so you won't see many examples of blatant disregard for worker safety. I know they exist, but to be honest, you don't need a book like this to help you see them.

Some of you may respond to a few of the stories by thinking:

- "That's so obvious. I would never do that."
- "Nobody would interpret that action in such a way."
- "That decision is a lot more complicated than you make it. Safety is just one of many things I have to manage."
- "That has nothing to do with safety."

- "We're in business to make money, you know!"
- "That's such a trivial example."

I encourage you to squelch those thoughts while reflecting on the stories. If you can't, you might want to put the book down right now. These stories provide an opportunity for you to see how people respond to the words leaders use and the actions they take. Put yourself in the leader's shoes, then the welder's shoes, or the project engineer's shoes. Stare in the mirror. I will help you, but you need to open your mind for it to be effective. If a story doesn't resonate with your role or your business, reframe it such that it does. Similar situations occur every day in most large businesses with high-risk work.

Some examples and my commentary may seem extreme, focusing only on the safety aspects of a decision or action. I realize many decisions are complex and driven by competing values: safety, cost, revenue, schedule, product quality, etc. However, you probably don't share all the facts, assumptions, and motivations behind your actions with the entire organization. Some employees will simplify their analysis of your decisions, i.e., cost vs. safety. If you say *Safety First*, then shouldn't safety be the predominant value influencing the decision?

You have many other people trying to influence your other business priorities; my job here is to hold the mirror for you…and whisper in your ear, *"Are you the safety leader you claim to be? If not, how can you?"*

The amount of work required to be an excellent safety leader may seem overwhelming. Don't get me wrong, leading excellence in safety is hard, probably the hardest part of your job. But the actions described here occur over years and decades, a few minutes here, a few hours there, and a full day next week. The best safety leaders integrate them into their work on other aspects of the business.

That being said, if Safety is Really First, shouldn't you invest more time and effort in it than anything else?

Summary

- A leader's commitment to safety is determined by how people in their organization interpret their words and actions. People are always watching for signals.
- SINCERITY plus CONSISTENCY equals CREDIBILITY.
- Mixed messages lead to loss of credibility and mixed performance. Mixed performance leads to injuries.
- Know the Talk, Walk the Talk, Hear the Talk.
- Look in the mirror and open your mind.

Personal reflection and learning

<u>Note</u>: Most people learn more by reflecting and practicing versus simply reading. At the end of each chapter, I include reflection questions and activities to help you deepen your understanding of the material, diagnose your own strengths and weaknesses, and generate ideas for improvement.

Before we start, use the following questions for a baseline self-assessment. When you finish reading the book, I suggest you come back to these questions to see how your answers changed.

- How do you describe yourself as a safety leader? You can use the questions at the beginning of this chapter to get you started, i.e., strong, weak, or good enough? consistent or inconsistent? proactive or reactive? caring or business-like?
- Why do you want to improve your safety leadership skills?
- Describe your safety vision, specifically.
- How consistent are your words and actions with this vision? How do you know?

- Describe the level of credibility you have as a safety leader. How do you know?

The examples in the book are intended to be mirrors for you to examine yourself. As you read them, ask yourself the following:

- Have I ever done that, or something similar?
- What safety messages did I send? How consistent were they with my vision?
- If I had been on the receiving end, what message would I have taken away?
- How would I change the words and/or actions to send a message that supports my vision?

2

What Do You Really Want?

What's behind the words?

- Goal Zero
- Target Zero
- No One Gets Hurt
- Safety is our Top Priority
- Zero Harm
- Safety is our Core Value
- Vision Zero
- Safety First
- Safety First, Safety Always
- Zero in on Safety
- Everyone Goes Home Safe
- No More Hurts

What are these anyway? Visions? Slogans? Targets? Programs? Words for a pretty poster?

Nearly every organization has them. It would be unconscionable not to, right?

And Zero what? Scratches, broken bones, fatalities?

They appear to say the same thing. But do they really? And even if it's clear to a leader, does it mean the same to everyone in the organization? This book is aimed at making it so.

The meaning is not in the words themselves. It begins in the mind of the leader of an organization. But the actual meaning is derived from the words they use and the actions they take. If different parts of the organization interpret them differently, how will they ever approach the vision?

Before a leader can consistently demonstrate sincere commitment to their safety vision, they need to understand specifically what it means to them. Let's look at some dimensions of these phrases and how a simple concept can be interpreted so widely.

What is harm?

The Vice President of Safety for the company was visiting my worksite. We saw her, or whoever was in the position at the time, once every four or five years. In a large conference room, she showed the safety performance data for the company: total recordable injuries, lost time injuries, and fatalities. The company was large, and one or more people had lost their lives at work every year for as long as I could remember. The recordable injury rate was stagnant at 1.2 per million hours, but no one had died at work in six months.

Then she made a comment that rocked my world. "We achieved Zero Harm for six months. You should be extremely proud, but also feel uneasy. We all know how quickly one event can change that. I beg you to plan your work as if tomorrow could be that day."

My mind spun, but I saw no visible reaction from the crowd. No shaking heads. No hands going up. No excessive squirming. Was I the only one? Granted, I was one of just a few HSE professionals in an audience of two hundred, but surely others must be confused. What about those eight lost time injuries on the chart? The broken foot, the concussion, the lost finger? I bet they felt harmed.

I prefer to take words literally, but have learned to read between the lines out of necessity. For over ten years, Zero Harm had meant zero recordable injuries of any kind to me, not just zero fatalities or zero lost time injuries. I reached that conclusion, starting with the words. Zero meant zero–clear enough. But zero what? It's implicit in the phrase. Zero injuries, right? But what type of injuries? Based on our target of zero recordable injuries and the reactions of some leaders to minor injuries, Zero Harm meant zero recordable injuries to me. Clearly, it meant zero fatalities to the VP.

But was it a target or an aspiration? Well, our annual target for recordable injuries was zero, but we never got close to that. Like most people in the organization, I knew this outcome would be extremely difficult to achieve in such a big company, so I treated it as an aspiration for my own sanity. The only way to get there was to have no injuries, one day at a time.

Walking back to my office with my manager, I expressed my surprise at the VP's comment on Zero Harm and asked what he thought. He said, "It made sense to me. At the corporate level, Zero means zero fatalities. At the lowest levels in the organization, it means zero recordables, and in the middle, it usually means zero lost time injuries." I discussed it with a few colleagues during the

following week and heard even different answers. After ten years,
Zero Harm was still not clear.

So what? Why does it matter how harm, hurt, or injury are defined? Don't we strive to prevent all injuries? In theory, yes. In practice, people make different decisions when trying to prevent hand lacerations versus falls from height. Many in the safety community feel that Heinrich's injury triangle is flawed, i.e., minor injuries don't necessarily lead to more severe injuries at decreasing rates. In other words, focusing on prevention of minor injuries has little influence on the likelihood of potentially fatal injuries. If a leader is aiming for zero fatalities year after year and the people on the front line are focusing on preventing ankle sprains today, how will they ever achieve the goal of zero fatalities?

In the safety vernacular, the main options for defining harm to people are:

- First aid
- Recordable injury
- Lost time injury, or days away from work case
- Significant injury or Fatality (SIF)
- Fatality

These are lagging indicators, i.e., the presence of bad outcomes. Zero has been aimed at them for decades. In recent times, more people are defining safety as the presence of controls or capability to manage hazards. These provide even more options to attach to zero, i.e., zero occasions with uncontrolled hazards, zero overdue inspections, etc.

I am not going to tell you what your safety vision or targets should be. That is for you and your organization to decide. Instead, in order to be an excellent safety leader, you need to know exactly what you want the organization to achieve so you can describe it and support it with your actions.

Before you read on, what does *harm* mean to you? What have you attached Zero to? What types of injuries are you focused on preventing? What is the phrase attached to your safety vision? What does it mean to you, specifically?

Vision or target?

Let's say you have chosen the phrase Target Zero to describe your safety vision to the organization. You dream of a company where no one is scratched or bruised, but you want to focus your initial efforts on preventing fatalities. So is Target Zero a vision, a goal, or a target? The label is important when describing it to the organization. Without further information, some people in an organization will interpret it as a vision, others as a target.

First, what do these terms mean? Many use them interchangeably, but they are quite different. For the purpose of this discussion:

- A vision is the desired future state. It is aspirational, inspirational, and conceptual.
- A goal or target is a specific result to be achieved within a specified timeframe, in support of moving toward the vision. Typically, targets are more specific and based on shorter time frames than goals. I will focus on targets in the discussion below.

The desired state described by a vision is generally far removed from the current state by design. The purpose of a vision is to inspire and align. The time period is usually undefined; in fact, you may never know when you are there. Visions provide high-level direction when employees face dilemmas between two or more conflicting priorities, but their specific actions may vary widely.

Targets and personal values drive most day-to-day behaviors. Targets are specific and time-bound, a recordable injury rate of 1.0 or less or one safety observation per person per week. They are often included in personal performance agreements, with positive and/or negative consequences attached. Therefore, specific work plans are developed to achieve them. "A goal without a plan is just a wish." (Antoine de Saint-Exupéry)

So, what's the problem? This seems clear so far.

Well, this happens…

> The tagline for the corporate vision was Target Zero. During a large construction project, the leadership team discussed and agreed upon specific safety goals, no lost time injuries (or worse) and no serious near misses (with the potential to kill). They tracked, reported, and investigated less severe injuries, but they focused on preventing serious incidents. In the previous project, two workers were seriously injured when they fell from height without fall protection. Hence, the team was most focused on working at height to prevent similar incidents or worse. It was one of hundreds of hazards, but the leadership spent twenty-five percent of their safety agenda on it.

> Their improvement plans were successful. They improved their performance significantly, but the threat was always there. Then they had a couple of minor hand injuries requiring sutures, making them recordable injuries. Upon investigation, the team found a few basic controls missing at some of their construction sites, so they corrected those deficiencies. However, they started getting more inquiries from above about hand injuries than working at height, vehicle collisions, and marine hazards. The total recordable incident rate was in the personal scorecards of their leaders, who were now in jeopardy of missing their target. Serious near misses were not in their scorecards. What were they to do?

The project team was receiving mixed messages. Should they continue focusing on serious injuries and near misses or switch their attention to preventing hand injuries? Their leaders wanted both, but only had annual targets for recordable injuries. The project team stayed the course, but in a less open culture, they may have shifted their focus after every incident...and perhaps someone would have fallen to their death.

For an organization to meet its targets and move toward its vision, everyone must be aligned on what they mean. Those who view Target Zero as a target of zero injuries of any type will probably focus their efforts on less serious injuries, since they are most likely. Those who view it as an aspiration and have targets related to more serious incidents will focus on hazards that can maim or kill. Don't get me wrong; no one wants anybody to get hurt. It is easy to say, "I want it all," but everyone has the same number of hours in a week. Their leaders expect them to complete a certain amount of work each week. They have budgets and deadlines. Where do they spend their limited time and resources to improve safety?

The solution seems simple. Just ask the boss or the boss' boss, "What exactly do you mean by Target Zero? Is it a target or an aspiration?"

Do you know how you would answer the question?

This question makes some leaders uncomfortable. Leaders who believe Target Zero is an aspiration may be reluctant to admit that some injuries may be inevitable this year. They may be vague or evasive in their descriptions so as not to admit so publicly. Those who believe Target Zero means no injuries at all may also be vague because they fear the organization will take it too literally and cause work to come to a halt or costs to skyrocket. Some just don't know what they really want. In any of these cases, people are likely to receive contradictory messages from their leaders about priorities in safety. They will revert to their own values or the wishes of their direct supervisor. Some will be aligned with their leader's intent, others will not. Excellent safety leaders describe specifically what they expect of the organization to avoid the pitfalls of ambiguity.

Use this question as a test. Would you rather have ten years with no recordable injuries and a fatality in the tenth year, OR ten years of a recordable injury rate of five per million hours with no life-altering injuries?

Don't say both; this is an either/or question. If you can't answer it, you're not ready to describe your safety vision and targets to your organization.

Priorities

Let's go back to the title of the book, *Safety First!* Or another popular safety slogan, *Safety is our top priority!* Is it really? Companies are in business to make money for owners or shareholders and to provide goods and services. Which companies are in business solely to *not get people hurt*? Even safety consultancies are in business to make money.

Supervisors and workers face competing priorities every day. In my opinion, this is the single biggest challenge in preventing injuries. No one wants themself or their colleagues to get hurt. But they also want to provide for their families, advance their career, learn new skills, and protect their job. The company also wants to lower cost, increase productivity, improve quality, increase profit, and deliver projects sooner. These goals are often contradictory and lead to compromises. Is the safest choice always made? If Safety was really First, wouldn't it be?

To make it even more difficult, *safest* is usually subjective. Some controls are universal: hard hats, safety glasses, gas testing before entering a confined space, fall protection when working at height, etc. But one can almost always make work safer. How safe is safe enough? Is the incremental benefit worth the extra time and cost needed to make a task safer? If it was simply a matter of desire, we would have solved this safety problem a long time ago.

Let's not get into the long-standing debate of whether safety is a priority or value. I think it is a value, but also believe it is not an important distinction for this dilemma. Many other business priorities are also connected to values: wealth, power, and family security. Whatever you call them, they lead to dilemmas. Since workers and organizations can have differing values, workers may make different decisions when faced with these dilemmas.

So, is Safety really First for you? Is it your top priority? You almost certainly will say so, but do you really mean it? Do your actions support it? If not, the organization will notice, and your future proposals or instructions related to safety will be met with skepticism.

- How much quality time do you spend on safety compared to other business priorities?
- How much resources (people, time, money) do you dedicate to improving safety versus improving production or delivering a project sooner?
- How often do you ask probing questions about safety vs. cost, production, project delivery?
- Will you always support a decision for the safest option, regardless of impact on other business priorities? If not, where do you draw the line?

These are good questions to test where safety fits in your priorities. If safety is your top priority, it will be most prominent in answers to the questions above–not just in how much time you devote to safety in town hall meetings.

Let's explore the first question in more detail. Does safety get 50% of your attention, making it clear to everyone it is your top priority? Or does it get 25%, on par with what you devote to production, quality, and cost? Or 10%, also leaving no doubt where safety fits on your agenda. You may be saying, "But I have leaders below me to lead safety." Do you really? If you are only spending 10% of your time on safety, do you really expect them to spend 50%? Most will focus their efforts on the other 90% on your list of priorities.

I once worked with a leader who actually thought about questions like this.

> We worked in a global project delivery organization. Most of the delivery was done through large specialist contractors around the world. A critical success factor for safe delivery of the work was to ensure that leaders of the company and the contractors were aligned on the safety vision and targets and how to achieve them. She could have delegated this to the project managers. She did hold them accountable for safety, but she also insisted on personally supporting the most significant of these engagements. With all the preparatory work and international travel involved, it was an enormous time commitment.
>
> While traveling to one of the large sites, she told me this one activity was the single biggest time commitment on her agenda every year. I had already noted her active participation in these engagements as supporting her safety vision and aggressive safety targets. But I was reassured when hearing this was a deliberate decision, based on careful planning of her actions to support her priorities for the organization.

What message do you think she sent to the project team? They heard the message back home, but also the message delivered to the contractor. They could check for consistency…or not. What message would the contractor's leadership receive when she traveled halfway around the world just to talk about safety? In cultures where hierarchy ruled, the top leader in the contractor company would feel obligated to participate, and their project team would take notice.

Let's consider another example where the messaging was not so clear.

Orders and revenues were declining. The leadership team agreed they must reduce costs in order to lower their pricing and still make a profit. Every quarter, the operations manager hosted an all-day meeting with each of his superintendents, the finance manager, and managers from the supporting organizations, such as maintenance, inspection, and logistics. Safety costs were integrated into the various department budgets, so the HSE Manager did not participate in the exercise.

Before the meeting, the finance manager sent each participant a detailed summary of each group's expenses for the previous three months, along with a template for entering their estimates for the next three months. The overall cost target was lower each quarter.

During the meeting, they scrutinized any increases in the actual or forecasted expenses over a certain threshold.

"Brenda, why did your chemical cost go up by 15% last quarter? Can you get the cost back down to where it was? If not, what other cost reductions have you identified to offset the increase?"

"Joe, you have consistently been under your estimate. I'm going to take 2% from your budget and give it to Brenda to help with the increased chemical cost."

And on it went, for eight hours.

Day to day, safety was given attention on par with cost, quality, and production, but the leadership team had no similar exercise for safety or any other priority. The closest equivalent for safety was the HSE Manager summarizing the past month's performance and a preview of upcoming safety events and issues during a twenty-minute time slot in the monthly leadership meetings.

Was safety really the top priority? What was the message to the leadership team? Did they identify potential safety implications of each cost reduction or reallocation?

The HSE Manager often wondered what might come from an equivalent effort focused on safety. One where each participant presented their incidents, trends, leading indicators, self-audit results, and safety observation trends, and others would inquire, challenge, and learn. One where the bar was raised each quarter.

***

Some of you may be saying, "I can't stay in business thinking like this. Work will grind to a halt." Then why are some of the most successful companies also the safest? I heard throughout my career that safe organizations are generally the highest performing overall. My experience made me a believer. Excellence is contagious. Credible safety leaders are credible leaders, period.

Your vision says Safety First. Mike Rowe said safety was third. What is it for you? Really? Let's try to find out throughout the rest of this book.

Summary

- Zero Harm, Goal Zero, and other such phrases mean many things to different people, from no scratches to no fatalities, from today to someday.
- A safety vision is aspirational; goals and targets provide more specific direction on day-to-day behaviors.
- A credible safety leader crafts their vision carefully, describes what they expect in detail, and reinforces it with consistent actions.

- Leaders who really mean *Safety First!* spend corresponding time and resources on it.

Personal reflection and learning

- What is your safety vision? How have you communicated its meaning to the organization? How specific was your communication?
- What are your safety targets? How well are they aligned with the vision?
- What kind of injury are you most focused on preventing?
- What is your safety slogan or tagline? What does it mean to you? How aligned is it with your vision? What does it mean to people in the organization?
- What are your business priorities, in order?
- How much time and effort do you spend on safety versus other business priorities?
- Discuss the safety messages being received from yourself with a representative sample of people in your organization. How consistent are they with yours and with each other?

3

Sincerity + Consistency = Credibility

Let's examine the simple safety leadership model introduced earlier in more detail.

SINCERITY plus CONSISTENCY equals CREDIBILITY. Credibility enables excellent safety performance that is sustainable.

Seems so simple, but how do you implement it in practice? The high-level traits of sincerity and consistency are supported by more specific traits which are more actionable. The following three chapters will explore each component of the model in detail.

The specific traits supporting sincerity and consistency are listed below. One could argue others should be included. In fact, my list varied through different drafts of the book. Each leader is naturally better at some than others. Workers place more value on some than others. There is no secret recipe.

A safety leader does not need to be strong in every trait. They can build on their strengths and be proficient enough in the others to prevent them from detracting from their messaging. If other traits work for you, don't abandon them. Use this information to supplement them and improve your credibility.

These traits are presented in the context of safety leadership, but this is just one aspect of business leadership. They cannot be considered in a vacuum. Some traits may be more important for safety leadership, but they are good leadership traits in general. For example, *caring* is an essential trait for safety leadership since safety is so personal, but it may be less essential for cost leadership. *Curiosity* is important to all aspects of leadership.

Many of the traits overlap and are interdependent. For instance, a leader can show genuine care for an employee by expressing curiosity about their family or personal interests. Persistence can be demonstrated by adjusting improvement plans based on feedback from employees. Giving this some forethought can reduce the time and effort to demonstrate the large number of traits. Perhaps you can connect the ones that are difficult for you to your strengths.

While writing this book, I found it difficult to place some traits in the correct category, sincerity or consistency. I quickly realized this was due to the overlap between the two. I settled on placing traits more inherent to the person under sincerity and those related to ongoing behaviors under consistency. You could try to justify why some should be in a different category, but please don't. You will see in the next few chapters how they work together. The model and categories are just an attempt to organize the complex nature of safety leadership.

Sincerity

No leader wants their people to get hurt, but what is their real motivation? Do they lead in safety for themself or for the people in the organization? This may seem trivial, but people can distinguish between the two, and it affects their behaviors. If a leader's efforts are clearly aimed at an employee's own safety, out of care, they are fully on board. If the efforts are intended to meet a corporate goal or boost the leader up the corporate ladder, not so much.

So how does a leader demonstrate sincerity?

Traits that demonstrate sincerity are:

- Caring
- Transparent (and Specific)
- Inspirational (and Bold)
- Competent
- Humble

Traits of a leader lacking sincerity are:

- Indifferent
- Self-centered
- Vague
- Arrogant
- Uninformed

Consistency

Many leaders can fake it in town hall meetings and Goal Zero award ceremonies. People know these orchestrated actions are easy; they are table stakes. What carries more weight are the leader's words and actions when their guard is down, when the main topic is not safety, and when business value drivers are in conflict. If the tough decisions are not consistent with their vision and carefully rehearsed words, then the orchestrated gestures are worthless to their credibility as a safety leader.

Traits that demonstrate consistency are:

- Engaged (and Visible)
- Curious
- Action-oriented
- Focused (and Organized)
- Persistent
- Appreciative
- Fair (but Firm)

Traits of a leader lacking consistency are:

- Distant
- Disorganized
- Disinterested
- Preferential / Unfair
- Passive
- Content

Credibility

This is where a leader's true commitment and hard work pay off. Once a leader has consistently demonstrated their sincere commitment to safety, people will trust them. Excellent safety performance requires continuous change, sometimes in big leaps, and change is difficult. People will give a leader with credibility the benefit of the doubt and follow them into the unknown. Without credibility, a leader's change proposals are met with skepticism or outright distrust.

A credible leader shouldn't think the war is won. Credibility is hard to earn, but easy to lose. A few decisions, new programs, or private remarks that are inconsistent with their vision will destroy hard-earned credibility.

Characteristics and outcomes of credibility are:

- Fragility
- Trust (and Respect)
- Support
- Approachability
- Innovation

Characteristics and outcomes of a lack of credibility are:

- Apathy
- Skepticism
- Distrust
- Stagnation

***

The next two chapters describe the specific traits listed under sincerity and consistency. For each trait, I provide examples of actions that demonstrate the trait, describe why it is important, and share stories that show the presence and/or absence of the trait. Finally, I provide reflection questions to help leaders determine how well they demonstrate the trait.

Many of the stories illustrate multiple traits, so don't focus solely on the trait described in the section in which the story is included. You may want to reread some of the stories once you've read through all the traits. Use these stories as mirrors to examine your own behaviors in similar situations, not just those of the leaders in the stories.

The lists of specific actions demonstrating a trait are based on experience and not an exhaustive list. Can you identify other actions you have used to demonstrate the trait? Some actions may help demonstrate multiple traits. For example, personally verifying the effectiveness of hazard controls at the frontline demonstrates being engaged (and visible) and curious.

Summary

- Sincerity + Consistency = Credibility.
- Sincerity and Consistency are supported by specific traits that are more actionable.
- Credibility gives leaders the permission to drive positive change.
- This model applies to other aspects of leadership as well.

Personal reflection and learning

- In what area of the business are you the strongest leader? Why? How can you use that strength to improve your safety leadership?
- What aspects of the model do you agree with? Disagree with? Why? How would you modify it?

4

Sincerity

This chapter describes the traits of a leader who demonstrates sincerity. The traits of an insincere leader are generally the opposite and are not covered specifically. However, the stories describe leaders who did and did not demonstrate the traits. Which is more representative of how you would act?

<u>Note</u>: Many of these traits have different meanings to different people and in different contexts. Hence, I have included synonyms and antonyms in italics to convey what I mean here.

Caring

Concerned, empathetic, and compassionate.

Not cold, disinterested, and insensitive.

These traits are not generally listed in order of importance, but caring is first on the list and first in importance–by far! Workers want to be cared for as people, husbands, mothers, and friends, not just as employees, and certainly not as numbers on a scorecard. They want to go home safely everyday so they can cook dinner, mow the grass, kick the ball with their daughter, go for a bike ride on the weekend, help their son with homework, or visit an ailing friend. Not because the corporate scorecard says the recordable injury rate should be less than 0.85. Not because a leader needs their safety ticket punched to get a bigger and better job.

For most leaders who truly care about their employees, this comes naturally. For those who don't, this trait is nearly impossible to fake. A caring leader sees no conflict between caring for workers and business success. Though they act out of genuine concern for their employees, they fundamentally believe satisfied and respected workers will be safer and more productive.

A few examples of how leaders can demonstrate care for employees are:

- Inquiring about their life outside of work: their families, interests, backgrounds, and dreams.
- Soliciting their input on working conditions, opportunities to improve their jobs or the organization, what messages you are sending about safety, etc.
- Acting on the feedback you receive.
- Ensuring injured workers receive medical care and other support to recover from their injuries, referring to the person by name vs. trade, job role, or IP (injured person), and staying in touch with them and their family. Privacy laws and policies may affect how much you can do.
- Spending time at the job site. Talking to the frontline. Acting on what you hear.
- Respecting and rewarding the contributions of all, regardless of job title, pay grade, or level of education.
- Helping employees maintain work/life balance. Setting an example yourself, or at least explaining your expectations of them.
- Being sensitive to personal struggles outside the workplace.

A caring leader gets to know people in the organization beyond their job title. Inquiring about a personal detail or two in one conversation can leave an enduring positive impression. Ask about their personal interests, their background, or their family. Find out what motivates and demotivates them: family, money, advancement. What exactly do they do in their role? How does it contribute to the bottom line?

The size of the organization will dictate how many people a leader can connect with and how deeply. If they don't regularly interact with some parts of the organization, they may need to create opportunities. The engagements are best done personally and informally, where the employees will feel more comfortable to share. The reach will be larger than those they engage with directly, as people will share their stories and impressions with others.

Be sensitive to those who don't like to mix their personal and work lives. Knowing and respecting their preference is a demonstration of care in itself. Conversations about an upcoming local event or their role may be a safer topic for them. To make them more comfortable, a leader can share something personal about themself.

Caring doesn't always involve pleasing workers with comforts and compliments. It may also take the form of tough love. Providing honest constructive feedback when someone disappoints you and the organization can be a gift as well. When done out of care, it is intended to help the employee learn and grow, not to threaten the worker or inflate the leader's ego.

***

Visit the worksites regularly. This recommendation will appear many times because these visits give a leader opportunities to demonstrate many traits and gather valuable insights into the people and the work. If done poorly, the visits can also highlight a leader's weaknesses. In addition to getting to know the workers through discussion, observe and inquire about the working conditions: temperature, cleanliness, access to bathrooms and break areas, availability of PPE, housekeeping, etc. What do they like about their job? What obstacles do they face? Take action on observations or common recommendations to improve working conditions.

STI would again be fabricating twenty-five small storage tanks for a new blending plant. The project manager, Matt, also managed the previous blending plant project and was determined that this one would be safer and faster.

During a kickoff meeting with STI leadership, Matt addressed the STI site manager. "Rick, during my visits on the previous project, I noticed field workers had no comfortable place to take their breaks or eat their lunch. During summer, workers ate their lunch and took naps in the shade of the construction trailer or even inside the rolled steel laying on the ground. They also had no place to wash their hands. One worker told me they weren't allowed in the construction trailer. How can we make their day more comfortable and their eating conditions more sanitary?"

Rick replied, "That's the way it's always been. We considered adding a lunch trailer, but they don't like to take their break with air conditioning because it's so hard to get going again in the heat."

Matt frowned. "I think it would be refreshing, but if they feel that way, there must be other options. How about covered picnic tables? And instead of port-a-potties, perhaps a trailer with toilets and running water, so they can clean up?"

"We can do that, but it will be more expensive, and I'm not sure it would result in better work. We scrutinized every penny to put together a good bid for you."

"How about we split the added cost? Would that work for you?"

Rick raised his eyebrows. "I guess so. You think it's that important?"

"Yes, I do. These men and women will determine how successful we are. We should do our best to take care of them. Why don't you survey the workers to understand what is important to them and present me with a recommendation and a quote?"

What a difference in demonstration of care between Matt and Rick. Rick demonstrated quickly that cost was a higher priority than worker comfort, even for the basics. Who wants to eat sitting on the dirt in the sun every day? Rick certainly didn't do that. What does that imply about the priority of safety?

Matt was disappointed. He awarded STI the contract, hoping they would make improvements after he had dropped a few hints during the previous project. But he opened up his wallet to make things right. His priorities were different. Both Rick's and Matt's teams took notice. Perhaps they would make better decisions on their own later in the project.

How well do you know the working conditions of your employees, especially the field workers? How would you feel if you were in their shoes? What will you do about it? What impact do you think improvements in working conditions will have on their safety and other aspects of their work? Though Matt was just acting out of care, chances are the frontline at STI will work more safely and deliver a higher quality product to his project.

***

A leader's response to incidents is perhaps the aspect of their behavior the organization watches most closely. Leaders show care by expressing genuine concern for the injured person, their family, and coworkers. They ensure the injured person receives the needed care and the conditions which led to the injury are corrected. They inquire about the person's condition, provide support to them and their family, and welcome them back to work when the time is right. Access to information on their condition may be limited by their own wishes or medical privacy restrictions. Looking for someone to blame and focusing on the classification of the incident will appear to be insensitive and selfish, i.e., the leader's reputation and safety metrics are more important.

If the hazard remains, take immediate action to protect others from harm, such as shutting down the activity or operation, excluding people from the area, and cleaning up debris or contamination. Reassure the organization by reinforcing your commitment to preventing harm and to learn from the incident.

Several examples of responses to incidents are included in this book. Let's start with a positive one.

Juan was still in the hospital after a terrible incident three weeks ago. He was painting structural supports, unaware that a welder was lighting his torch around the corner. He was burned severely, had two surgeries already, and had at least one more to go. And to make it even worse, his family lived over 300 miles away, restricting their visits to weekends.

Rodrigo, the site manager of the small fabrication yard, visited him every few days, but wanted to do more. He asked the corporate office for money to help Juan's family with travel costs, but they were worried about the legal implications. Rodrigo shared the disappointing news with his leadership team.

A week later, Steve, the coatings inspector, came to Juan's office and tossed a fat envelope on his desk. Juan opened the flap and exclaimed, "What the heck is this?"

"We were disappointed after the meeting last week, and a few folks asked me if we could make our own donation to Juan's family. Word spread quickly, and $2100 poured in this week. Even the client team members chipped in. I know the head office said they couldn't do anything, but this is from us, not them."

Steve hesitated for a few seconds before replying. "It's a little risky, particularly for me, but it's the right thing to do. How can I not support everyone's generosity? Here's another $100 to add to the kitty."

"Great. Thanks for contributing."

"I'm planning to visit Juan on Saturday, and his wife is always there. Why don't you and a few of your teammates come with me to present your gift? Include the client lead as well. I'll take the heat if corporate calls."

Juan's teammates showed their care and generosity. That wasn't surprising. They worked with him every day; some were his friends. Rodrigo had a dilemma, but chose care over potential admonishment from the corporate office and risk to his career. Imagine the message he would have sent Steve and the other contributors if he had tried to prevent the donation.

What is your first reaction when someone gets hurt? Or involved in a serious near miss? I found that nearly all leaders inquire about the status of the injured person and whether the hazard has been removed or isolated. The most caring ones continue to monitor the person's recovery and seek to learn from the incident. The others quickly move on to discussions about case management and the classification of the incident, sometimes even blame.

Which leader are you? Your organization already knows. Which do you want to be?

***

Safety professionals frequently talk about situational awareness in the context of workplace safety. Leaders can demonstrate care by showing situational awareness on a personal level as well. They can acknowledge unusual behaviors or emotions and ask if they can help. The conversation may not go far, but the offer will be appreciated, and the person may need the help. Sometimes a leader can help by holding off on significant requests and/or reallocating work while the person is struggling.

> Jane was excited about her first new role in twelve years. She was confined to a large chemical plant during that time, but was now in a role supporting plants around the world. She had to deal with multiple time zones. During her second meeting with her new manager, Claudia, she described a new virtual team she had joined, composed of members from Singapore, Qatar, Germany, and her in the United States.
>
> "They meet every month at midnight, our time, way past my bedtime, but I need to catch up and make a good first impression. The tough part is that I wake up early to participate in calls with my colleagues in Europe," Jane said.
>
> "Glad to hear you are making some good contacts. Before we discuss those meetings, can you tell me a little about your family? This new role may be an adjustment for them as well."
>
> "Sure. My husband, Peter, is an accountant and does a lot of financial auditing, so he travels a lot. We have one daughter, Ashley. She's in high school and pretty much self sufficient. She's thinking of becoming a doctor, and we're both excited about that."

"Thanks for sharing. My son is a medical resident. Maybe we can talk later about what he's learned over the past few years and put them in touch?"

"Great idea. I don't think she's thought through all the implications yet."

"Back to those meetings…I've been in a global role for eight years. You'll be asked to join calls at all hours. I recommend you set boundaries on your working hours and stick with them. Working from home gives you more flexibility, but you shouldn't let work invade your personal time too much."

"Thanks for the advice and understanding. I'll try it for a couple of months. I've never experienced a role like this. At the plant, I got calls in the middle of the night, but only for emergencies, and the adrenaline kicked in quickly."

"OK, but think further about my advice, and let's discuss it again in a month."

The evening before the first call, Jane struggled to stay awake, even at the beginning of the call. She set her alarm and took a short nap, which just made her disoriented. During the call, she couldn't follow the discussion, and her responses were delayed and disjointed. After the call, she worried about the poor first impression she made on the team and recalled Claudia's advice. She would find another way to stay connected, perhaps by calling team members separately at more reasonable times.

During her next meeting with Claudia, Jane said, "You were right. I'm sorry I didn't follow your advice straight away. The first call ruined my evening and the following day. I'll find another way to stay in touch." She changed subjects. "By the way, the people at the Singapore plant requested I visit to review their program. I'd like to swing down to Australia to visit another plant to save a long plane ride."

Claudia went into coaching mode again. "That's great. You're gaining valuable credibility in your new role already. Our team has a rule of thumb that you shouldn't be away from home for more than two weeks at a time. Those long trips wear you out and can be hard on your family. Please keep that in mind when planning the trip."

"You were right last time, so I'll take your advice this time. I'm eager to get off to a good start, but thanks for advising me to pace myself."

Don't underestimate the power of what seems like such a trivial example. The message was clear. Claudia would only have this discussion with Jane if she cared about her as a person. She knew Jane would be eager to make a great first impression, but she also recognized that each person was different and gave her the space to find her way. Fortunately, Jane found it quickly. It is tough to be open with a new manager, but Claudia quickly earned Jane's trust. On the surface, it may seem Claudia was sacrificing productivity from her team, but Jane will be much more likely to step up when a crisis arises.

Do you know how work is affecting other aspects of an employee's life? How do you find out? How do you accommodate employees' unique styles of work? How good of an example do you set yourself?

Key actions

- **Show care for people, not numbers**

Personal reflection and learning

- What settings and approaches do you use to get to know your employees?
- How well do you know what dilemmas, discomforts, and stresses your workers are feeling? How do you find out? What is your reaction to such feedback?
- How well is your demonstration of care balanced between those closer to you and further from you, in the office and in the field, one site versus another?
- What obstacles restrict you from showing more care? How can you overcome them?
- What advice do you give employees on the balance between work and home life?
- How do you react when someone gets hurt? How about serious near misses?
- How do you ensure the conditions leading to an incident have been corrected immediately?
- How involved do you get in case management and incident classification?
- How else do you show care for your employees?

Transparent (and Specific)

Open and honest, but also clear and specific.

Not deceitful, evasive, and vague.

Closely related to Engaged (and Visible).

Leaders can be Transparent and Specific by:

- Clearly communicating their safety vision to the entire organization. Safety First is not clear enough. If you use the word *Zero, Zero What?*
- Developing specific targets consistent with the vision.
- Helping people solve dilemmas involving application of the safety vision in their roles.
- Helping other leaders develop specific goals for their teams to make progress toward the vision.
- Following the rules themselves.
- Being open about their motivations in key decisions, especially those that may seem contradictory to the safety vision on the surface.
- Openly sharing and acting on lessons learned from incidents.
- Establishing clear accountabilities for other leaders in the organization, especially where boundaries are fuzzy.

This trait has several components, so let's break it down. The definition of transparent includes the words open and honest. In other words, tell the truth…the whole truth. A leader that hides or misrepresents their motivations and aspirations on safety won't be trusted. They will be unable to build credibility. Examples of dishonest leadership behaviors include under-classifying incidents, not following safety rules themselves, setting targets they know are unachievable, ignoring reported hazards, and committing to improvement actions with no intent to follow through.

Doing the right thing sometimes involves employees taking risks and being uncomfortable, for example, stopping work, providing feedback on a poor supervisor, reporting bad news, and implementing change. All are critical to improvement, but will employees take the risk if they don't know what their leader stands for, or worse yet, fear retaliation?

Openness and honesty are critical, but not enough. People also want to hear specifics. Details help convince people a leader is being genuine. Relevant details are harder to fake than vague slogans. Details indicate a leader has thought deeply about how the organization will move toward the vision.

A leader can begin demonstrating transparency by explaining specifically what the safety vision and targets mean to them and what they expect from the organization. I covered a lot of this in Chapter 2. What does Zero Harm mean? What type of harm? Is it a target or an aspiration? Vague descriptions and evasive answers breed distrust and skepticism. How committed to safety is a leader who cannot describe their expectations in detail?

Visions and targets are necessary, but how will the organization achieve them? Share specific action plans to achieve the targets for the year. From there, each department and person can figure out how they can contribute. If everyone moves in the same direction, the organization is more likely to meet its targets.

While auditing, I usually asked the top leader at a facility, "What do you personally do to improve safety performance?" Answers such as "I address it in the town hall meetings" or "I do a safety walk once a month" don't convey a lot of personal commitment. If I could sense it in five minutes, their staff already knew it. That set the tone for the entire audit. If they couldn't identify specific and material actions, how well did they really know their vision and how to get there?

✳✳✳

Let's consider a few examples in more detail. First, let's see how a leader can be specific about safety targets and inspire their team to develop action plans to exceed them.

> Bjorn was the project manager of the largest project in the region. His leadership team was the best of the best, and all the senior

leaders would be watching the project. The leadership style he had honed over a couple of decades was to be specific about expected performance, hold his leaders accountable, and inspire them to deliver even better results. In November of each year, he gathered his leadership team for two full days away from the endless emails and standing meetings. The first few hours were dedicated to safety.

The first item on the agenda was to review safety performance for the current year and set the safety targets for the following year. The broader organization had a Zero Harm vision and a recordable injury target of 1.0 (per million hours). Some project managers would view those as specific enough, but not Bjorn. He had his own views, but what should the project team be striving for, a recordable injury rate of 1.0, 0.5, or 0.0, zero LTI's, zero fatalities, zero high potential incidents, no falls from height, all of that? The target for no falls was on the menu because of the rash of serious near misses earlier in the year and the subsequent focus on minimizing that risk. Bjorn felt he needed this clarity so his team could develop action plans specific enough to impact performance.

He wanted open discussion of the options, but being such a sensitive subject, he put a stake in the ground to begin the discussion. He proposed that Zero Harm for their project for the next year would be defined as zero lost time incidents. Since working at height was still a threat, he also proposed a target of zero high potential falls, i.e., those with the potential to kill.

With such an experienced team, Bjorn didn't have to wait long for the discussion to begin. The construction manager, who was responsible for most of the safety exposure in the upcoming year, chimed in first. "I support the LTI target. We have focused on reducing the more serious injuries from the beginning of the project.

I don't think it's feasible to expect zero recordable injuries this year. We have so many hazards in our work, and can't focus on everything."

The safety manager asked, "Don't we need a recordable injury target since your boss has set a target of less than one?"

The project manager replied, "I have talked to him already. He supports our focus on high potential incidents, and has left it up to us."

The discussion continued for another fifteen minutes, with Bjorn encouraging views from the quieter team members. The team supported the proposed LTI target and decided not to set a project-specific recordable injury target.

They moved on to the target for falls. The quality manager opened the discussion with a challenge. "I am afraid a target of zero will set us up for failure. We had three in the first half of the year. Granted, we've only had one in the second half, but getting to zero will be difficult. What if we have one in February? I'm afraid people will lose motivation since the target is no longer achievable."

"But shouldn't we strive for zero? How else will we get there?" said the safety manager.

After Bjorn solicited a few more views, they agreed to a target of one (or less) high potential falls from height. They also adopted aspirations of zero recordable injuries and zero high potential falls from height. The targets were their commitments to each other and to the rest of the organization, but Bjorn wanted to inspire his high-powered team to do even better.

> His parting remarks were, "This is not just a numbers game. We had this discussion so we would know what to focus our actions on. I want each of you to identify one specific action you and your teams will take to achieve each of these targets. Make it specific to your work. Send copies to the safety manager so she can compile them and share with all of us for learning."

It would have been easier for Bjorn to adopt the broader organization's Zero Harm vision and target, e.g., no recordable injuries. But he and his team were experienced and understood their largest risks during the next year. Bjorn sought his team's input and adjusted accordingly to create joint and personal ownership. He insisted on specific actions aimed at achieving the targets. Imagine how those actions would look if they had adopted the generic Zero Harm vision. Some teams would work on hand injuries, others on working at height, and others on mental health. Instead, they would all be working to improve the controls related to the highest risks in their scopes of work, and fall prevention and protection specifically.

The leadership team left the room energized but uncomfortable. The targets would be difficult to achieve, and they would be held accountable for their results. Bjorn would be tested throughout the year, whenever someone needed a few sutures or had a metal shaving removed from their eye. Of course, no one wanted those incidents, and they would have controls in place for those hazards, but their improvement efforts would be focused on the big stuff. His team would watch his reactions closely, and those of his boss.

How much clarity do you provide on your safety vision and targets? How well defined and aligned are the actions of your team members? Will the actions truly enable the organization to meet its target?

***

Now let's examine another aspect of transparency, honesty. People are watching their leaders' safety-related behaviors all the time, but the scrutiny is generally highest after an incident. People are curious because they care about the injured person(s) and don't want the incident to reoccur. Generic descriptions of the incident and delayed information build suspicion, so share accurate information as quickly as possible. Focusing too much on the classification of the incident may be viewed as uncaring, i.e., prioritizing numbers over people. Under-classification of incidents will be viewed as dishonest. The leader loses credibility, and the targets and vision lose meaning.

Derek yelled, "Terry, watch out!" as he noticed a section of grating missing just in front of his colleague. When he had noticed red barricade tape on the ground, he was looking for the hazard the tape was intended to isolate. The warning was too late. Terry disappeared through the open hole as he turned to look back at Derek. Through the grating, Derek saw Terry's hip bounce off a twelve-inch pipe before falling to the grating of the deck below and landing on his side. Derek called the control room on his radio for emergency assistance.

Terry was fortunate his head didn't strike one of the many obstacles between the decks, but he did break his pelvis and an arm. Later that day, he underwent surgery to repair his pelvis. The doctors originally believed the arm would heal with immobilization, but the following day, he was in surgery again to stabilize the break in his arm after seeing a specialist.

The Operations manager answered the phone the following day. "Hello, Steve here."

"Hi Steve," said Stephanie, the HSE Manager. "I wanted to follow-up on our discussion yesterday and check how Terry is doing."

"Derek talked to him about an hour ago. Terry will be OK after a lot of rehab. He had another surgery today for his arm. They were going to send him home tonight, but he lives alone, so they didn't want to send him home quite yet. They are hoping he can find someone to stay with him."

"Oh, sorry to hear about the second surgery, but good to know he will recover. What a scary fall! Has the grating been repaired yet?"

"Yeah. We replaced the grating this afternoon when the materials were delivered."

"Do you need any help from my team for the investigation? Since this is a lost time injury, the investigation needs to be thorough."

After an unusual silence, Stephanie asked, "Steve, are you still there?"

"Yes, yes. We'll investigate it thoroughly so it doesn't happen again, but this isn't an LTI."

Stephanie was caught off-guard and paused. "Steve, Terry has already missed a day of work, and I don't see how he will be back for the next few days. I think it's clearly an LTI."

"Yeah, but if the doctors had realized surgery was needed on the arm yesterday, he could have had both done at once, and maybe he could have been at work today, in a restricted capacity. My safety technician said he has seen that done before."

Steve paused, then continued. "Listen Stephanie, the site hasn't had an LTI in two years, and my performance contract says zero LTI.

The team will be disappointed if we break our LTI-free streak. If there is any logic to support a lower classification, I'm going to use it."

"Steve, it doesn't matter. If he misses a day of work, it is an LTI. It's as simple as that. I'll talk to our boss about it."

"I already have, and he was fine with it. Hey, I need to run to the work permit meeting. Talk to you later."

Stephanie walked down to Frank's office. Frank was the manager of the site. "Hi Frank. I just talked to Steve, and he told me Terry's prognosis is good.

Frank replied, "Good to hear he will be OK. I don't want anyone to go through that again."

"Did you talk to Steve about the classification?"

"Yes, he told me it's a restricted work case."

"But Frank, he's still in the hospital."

"He told me that Terry could have been back at work the next day if the doctors had been more coordinated."

"The classification doesn't work that way. If he misses work due to the injury, it's an LTI. The detailed circumstances don't matter. It sounds like Steve will be investigating it thoroughly either way. That's good, but consider how the rest of the plant will view the classification. Some will feel this is being swept under the rug."

"OK. I'll talk to Steve. But people in the plant won't react like that. Thanks for the advice."

The tough life of an HSE professional, holding up the mirror and being the bearer of bad news. On the positive side, it appeared everyone was concerned about Terry and determined to prevent recurrence. A thorough investigation was in the works. Stephanie represented her discipline well and stood her ground. She prevailed in the end and prevented most of the damage to Steve and Frank's credibility. However, their credibility with her was significantly lower now. Unfortunately, some other members of the leadership team blamed her for the end of the LTI-free streak, not the unmarked open hole Terry fell into.

Despite Steve's apparent care for Terry's recovery, he was looking for reasons not to classify the incident as an LTI. It would end the two-year streak. Staff bonuses might be impacted. His future job opportunities might be affected. Frank may have had similar concerns, but he appeared to be more distant and indifferent. Some leaders believe these attempts to under-classify an incident won't be noticed by their organization, or they won't care. Nothing could be further from the truth, especially for those closest to the incident. How do you think Derek would react? He saw the fall and heard the screams. He saw Terry's grimaces when visiting him in the hospital. All it takes is for a few people to notice and start asking questions, making assumptions, and spreading the word.

If Stephanie had not prevailed, how would Frank have responded to questions in the next town-hall meeting? Steve and Frank would be reluctant to share details about the incident as it might expose the blatant under-classification. Steve and Frank would be viewed as dishonest, evasive, and showing more concern for the numbers than for Terry. The LTI-free streak would no longer be a source of pride, but a reminder of the steps their leaders would take to make performance look good on paper.

***

Finally, let's consider being open with the organization. The safety messaging from decisions made about dilemmas can be difficult to interpret for those not in the decision-making process. Workers rarely have the same information the leader has when making the decision. On the surface, the leader may appear to choose an option that is not the safest, but he or she may have seen other potential safety impacts. To prevent misinterpretations, a leader can acknowledge the dilemma, the conflicting values, and explain why they made the decision. They may not convince everyone they made the right choice, but nearly everybody will appreciate the transparency.

> Here we go again! The third staff reduction in five years at the plant. The message from the corporate office was clear; reduce costs, and people cost a lot. But this one was different. The refinery manager openly acknowledged that if the staff count changed, the work had to change. A base amount of activities were mandatory to keep the plant running safely. No one knew what the number was, but they all felt the last round of reductions had put them there.
>
> "Every manager will form teams to list all their activities. Each one will be scrutinized. Is it necessary to keep the plant running? Is it necessary to keep people safe, prevent leaks and spills, and comply with our permits? If we stop it, what would happen? And I want every manager to include at least two people from other departments they interact with the most to provide additional perspectives."

Refreshing or too good to be true? Was this just a ploy to head off the inevitable questions about prioritizing cost over safety? Or would the analysis really affect the outcome? This particular plant manager had already proved that he truly cared about safety, the environment, and their impact on the surrounding community. He had credibility. So the department managers diligently followed his request. But what if he didn't have credibility? Would the managers have just gone through the motions, knowing the answer was in the back of the book? What if it wasn't? What if he was sincere this time? The outcome would be based on poor information. Credibility is a powerful force, both good and bad.

Key actions

- **Be specific about your vision and expectations**
- **Follow the rules yourself**
- **Share the true motivations behind key decisions, especially dilemmas**

Personal reflection and learning

- Which aspects of your safety vision do you struggle to explain to others?
- What specific examples have you shared with employees on what the safety vision and targets mean to you?
- How aligned are employees' safety goals with the safety vision and targets?
- How often do you seek loopholes versus meeting the intent of the rules?
- How do you acknowledge the potential safety consequences of broader business decisions, especially when there are conflicts between safety and other business drivers?
- How else do you demonstrate transparency?

Inspirational (and Bold)

Motivating, challenging, courageous, and ambitious.

Not ordinary, routine, weak, and content.

Closely related to Action-oriented.

Leaders can inspire their organizations to achieve excellence in safety by:

- Helping them understand that the vision is worthwhile and achievable and that they can make a difference.
- Setting bold targets and taking bold actions to achieve significant improvements.
- Backing their bold commitments with the resources needed and their own time and support.
- Being patient and consistent, even in the face of inevitable setbacks.
- Recognizing progress along the journey.

Sending everyone home unharmed every day is hard work. I believe it is the most difficult task most leaders have, especially those with significant safety exposure in their business. Hazards abound, people make mistakes, and the stakes are high–life and death! An organization must display excellence in many ways to approach this lofty aspiration: comply with hundreds of safety regulations, identify new risks, manage changes, communicate clearly, perform quality work, etc. A safety leader must inspire this excellence in their organization. Without excellence, Zero Harm, whatever that means to them, is impossible over the long term.

Inspiring excellence begins by describing the desired outcome and why it is important to achieve. The outcome should be a stretch, maybe even a dream. Though the journey may be long and difficult, the leader should build confidence that it can be done, one day at a time.

In this context, the desired outcome is a safe workplace, however you have defined it. An organization cannot be excellent in all aspects of their business. They must prioritize and focus their efforts. For example, quality (including safety), cost, and schedule are the three of the pillars of project management. Many project managers will tell you, "pick two of the three–you can't have them all." If Safety is really First, then shouldn't it be on the top of the list?

A clear description of the desired outcome is only the beginning of the inspiration; actions operationalize it. A leader must put their money and time where their mouth is. Action will be discussed in more detail under *Consistency*. Suffice it to say for now that a mix of bold actions, frequent and genuine messaging, and continuous improvement are required.

Many leaders will attempt to implement small changes to their safety management systems every year. They may not know what else to do. They may fear bigger changes will interfere with other business priorities or worry about unintended negative impacts on safety. Small changes usually result in performance improvements just as small as the changes. Little ventured, little gained. Much bolder actions are needed to drive improvements that are significant and sustainable. Most organizations have enough headroom between their current state and Zero Harm to justify such bold actions. Boldness inspires. If a leader takes on such a big change and the risk associated with it, people are more likely to follow them.

Carefully balance recognition of progress with the quest for even better ideas and results. People can become frustrated if what they do is never good enough. Genuine recognition can motivate people to continue moving forward on the never-ending journey. This can become self-sustaining when the hard work shows in the results, e.g., fewer injuries, less severe injuries, and better risk assessments. A safer workplace is the reward. But until that happens, a leader needs to provide recognition to boost confidence and motivation.

***

Now, let's look at a bold promise that was so inspiring the leader never had to implement the corresponding action. This example is about environmental leadership, but is relevant here since Health, Safety, and Environment (HSE) are commonly managed together.

Alexis was the environmental engineer at a plant. Her primary role was to help the operators of the wastewater treatment plant ensure the effluent always complied with the wastewater discharge permit. One of the plant's HSE targets every year was zero permit exceedances (there's that word again). They achieved the target in most years, and when they didn't, it wasn't a result of the chemicals coming from the process units; it was the dirt in stormwater.

The plant was built on clay. It didn't rain hard often, but when it did, the water picked up tiny clay particles, threatening compliance with the permit limit for total suspended solids content. Similar solids were carried down the local rivers, but that didn't matter; it was in their permit. The only practical solution was to impound the water to allow the clay particles to settle out. After heavy rainstorms, the operations team rented diesel irrigation pumps and hoses to pump water into holding areas, such as a diversion basin built for that purpose and tank dikes for tanks no longer being used. The overtime and rental costs were never questioned.

During the worst storm of Alexis' tenure, the available storage areas were full and more rain was forecast. She didn't know where to put it. She rarely attended the daily leadership planning meeting, but she made an appearance today. The goal of zero permit exceedances was in jeopardy. She had exhausted the options she controlled. The only one left was to reduce the flow coming from the process areas, which would slow down production. She briefed the leaders on the predicament. They knew something was brewing, but her presence confirmed the dire situation. The plant manager, Dick, didn't hesitate. "You let me know when, and we'll slow things down."

Alexis inhaled sharply with surprise. She hoped it wasn't too obvious. She knew the target of zero was worth some overtime and rental fees, but that was small change compared to slowing down production. She now knew zero really meant zero!

Dick's response was immediate, specific, and clear. Not "We'll think about it" or "How long can we wait?" Without his strong credibility built over years of similar statements and actions, his team may have called his bluff. But they knew he was serious.

It would have been easy for Alexis to ask for the slowdown while she was there. However, she was now even more determined to solve the problem on the back end, and she did. She never had to ask for the slowdown. She was also motivated to prevent future close calls. Perhaps she would put together a proposal for a permanent pumping system and more storage areas. She had considered doing so before, but assumed the extra cost would never be approved. It now appeared that was a poor assumption.

How do you think the operations managers reacted? How would you have reacted? I suspect some of them encouraged their teams to find ways to minimize their discharges without impacting production. Going forward, they would pay more attention to the discharges they sent to the wastewater plant. They were inspired to be excellent. They were now convinced the target for zero permit exceedances was an expectation versus an aspiration.

How do you think this impacted organizational support for the plant manager's other targets and improvement plans? What if he had told Alexis to just deal with it? Are you prepared to choose safety (or environment) over production, cost, or schedule? How dire does it need to be? That may tell you what zero really means to you.

***

Next, let's consider an example showing a couple of bold actions, one by the line leader and one by the HSE manager. Together, they solved a crisis, but also led to longer-term improvement.

Simon was the HSE manager on a multi-billion dollar project to build a new chemical plant. The production modules were being built in a fabrication yard in the Far East, which had been used on past projects. The project team knew dropped objects would be one of the top risks. Others included falls from height and confined space entry. Safety performance in previous projects was OK, but they aimed to do better, i.e., zero serious injuries. Simon's team of safety coaches advised the contractor's supervisors and frontline workers on safe work practices.

Unlike previous projects, the project team developed a robust plan to prevent dropped objects. At least, they thought it was robust. They trained the workforce, purchased lanyards of all sizes and types, draped netting on guardrails up and down the structures, and provided miles of red barricade tape for exclusion zones. They were confident.

In the first nine months, they had no serious dropped objects, further bolstering their confidence. But the work was still near the ground. Then, in just one week, three objects with the potential to injure someone seriously fell out of the structures, including one that grazed the arm of a scaffolder. He was shaken, and the project team was scared. After the first incident, they started an investigation. Before they even finished it, they began another. After the third, the project manager, Janice, called time-out and stopped work for two weeks.

The investigations revealed that even though they had the materials and training to prevent dropped objects, the frontline workers were not implementing the controls consistently or effectively. Of course, the frontline wasn't the root cause. Project and contractor leadership had grown to tolerate uncovered gaps in work surfaces, three out of four tools being secured with lanyards, and guardrail netting fluttering in the breeze. A new norm had been established, and they were paying the price. The exposure would only increase as the work moved higher.

They used the time-out for two things. First, they patched up the gaps on work platforms, purchased more tool lanyards, and established no-go zones in high-risk areas. But these quick fixes would not last unless they changed their systems and behaviors. They decided to measure the effectiveness of the controls every day at the frontline, where it mattered most. They are the ones exposed; they put the controls in place. The HSE Manager changed the role of his team from coaches to inspectors. They still advised and intervened, but they observed and recorded the effectiveness of the controls at each worksite. Not after the intervention, but as they arrived at the worksite.

Using data on the condition of controls at the frontline to hold the leaders accountable made the difference, not the extra lanyards they bought or the pep talks by the big boss. Their leadership and the system change made the difference. They reduced the number of high risk dropped objects by over 80 percent in one year. Considering the previous rash of incidents, they may have saved a life or two.

The first bold move should be obvious, shutting down work for two weeks. It wasn't an easy decision. The two-week delay could have meant two weeks of deferred production, a lot of money and a lot of broken promises. Clearly, safety was more important than schedule and cost to Janice. The contractor was shocked. Clients had shut projects down for a day for a safety stand down, but never a week, much less two. The first day or two had little impact since it was within their experience. After about five days, they realized Janice was different, and they would have to adapt. The project team also noticed. They now knew that if they didn't get safety right, their careful schedule planning and cost management would be for naught.

The other bold move continued to pay off every single day for three years. By implementing the verification scheme for controls at the frontline, project leaders at all levels received weekly feedback on the strength of the dropped object controls. They took actions to rectify deficiencies and to improve continuously. This may not seem like a bold move on the surface, but the contractor initially viewed it as policing, and the safety team had their daily routines upended. When the leaders finally acted on the data and witnessed the improvement, they all became supporters.

The HSE Manager saw his former boss halfway through the improvement journey. He knew the story and recognized the team for turning things around. But he asked one question that had a lasting impact, "Why did you expect a different result when you didn't really change anything from the past projects?" It stung for an instant, like a bandage ripping the hair off of his forearm. But he was right. Yes, they had a better dropped object prevention plan, but they changed nothing about the cultural differences between the two organizations. They had just tweaked things. In this case, the tweaks resulted in slightly worse performance, but the best they should have expected was marginally better performance. Their ambition was much higher. Small changes to inputs usually lead to small changes in outputs. Bold changes are needed for step jumps in performance, like achieving Zero.

****

In some ways, making reactive bold moves like the one above is easier since the case for action is clear. However, the changes are disruptive and can jeopardize other business objectives. An excellent safety leader makes changes proactively so they can be carefully planned and implemented.

Other examples of bold moves include:

- Reducing exposure to hazards by replacing or eliminating equipment, using better means of work access, and changing chemical formulations for feedstocks or manufactured products.
- Changing a contractor after attempts to improve performance of the existing one are unsuccessful.
- Purchasing and maintaining supplies of different types of high quality protective equipment.
- Selecting a contractor with the strongest safety program and performance instead of the lowest bidder.
- Replacing a senior leader whose behaviors do not support the safety vision with a proven safety leader.

Challenge yourself. Let's assume safety truly is your top priority. What is your second priority? If you were struggling to succeed in that area, what would you do? Are those actions bolder than what you are doing for safety? If you have successfully made bold changes in other areas, how can you use a similar approach for safety? Would you replace a project manager whose project continues to miss milestone after milestone despite repeated coaching and support? What about one with an injury rate twice your target or one not supporting your safety improvement agenda but hitting every milestone? Look at yourself in the mirror.

***

Leaders won't achieve safety excellence without choosing business partners with similar ambitions. If safety is your first priority, wouldn't you want to work with those who are the safest in the market, while still delivering the quality product or service you expect? Would you be willing to pay more? Why not?

Franco, a safety advisor, met Jose, a maintenance supervisor, at the cooling tower. Franco had called fifteen minutes earlier to discuss an incident, but he insisted they do so at the site. Jose was standing four feet away from a scaffold tube sticking out of the wet ground near the foundation of the tower.

"Hi Franco. Now you know why I wanted you to come out here. No one was hurt, but one of the scaffolders' ground helpers was standing here when this pole fell. This new scaffolding contractor is going to hurt someone bad if we don't do something, and it may not be one of their own."

Franco shook his head. "Oh boy, here we go again. What is this, the third dropped object since they started working on site last month?"

"Yeah, and we've seen several of the scaffolders not wearing their fall protection equipment. I don't understand why we had to switch contractors. The previous one built good scaffolds and had a great safety culture. They were also flexible with our ever-changing priorities. I guess this new company is cheaper."

"I don't know. I wasn't involved in that, but you're probably right. If the people who make those decisions could only see the impacts at the frontline. They seem to believe we can spend some of the savings to make them safer, but we never get any of that money, and it's not so simple."

Jose said, "You got that right."

"Thanks for calling me out here. I'll volunteer to lead the investigation. We'll dig much deeper than blaming the guy who dropped the tube. Maybe it will shed light on the contractor selection process."

"Thanks, Franco. They keep saying Safety First, but this new contractor doesn't convince me of that. Makes me think it's just a catch phrase and nothing else."

"You'll probably be interviewed, so don't hold back."

Jose grinned.

Franco and Jose had become skeptical of their leader's Safety First vision. The change from a contractor providing both a quality service and excellent safe work practices to one with much worse safety practices showed them how little their leader valued safety. How inspired will they be to influence the new contractor to improve their safety practices? They'll probably try, based on their personal commitment, but they are unlikely to be successful without support from the rest of the organization. They'll spend their time helping the contractor catch up versus moving the rest of the organization forward.

Since the new contractor probably knows how their work practices compare with their competitor, the maintenance manager just showed them that worse safety practices were acceptable by awarding them the contract. How inspired will they be to improve? They know they have to keep their costs down to keep the contract.

Many organizations work with business partners, including contractors who may perform the most hazardous work at a facility. Leaders should insist on safety excellence in these partners and make it a key factor in the selection process. Safety performance and capability are generally included as one of many contractor selection criteria. However, how much weight is given to safety in the selection process, 10%, 50%, 90%? How accurate is the information used to assess their safety capability? Does it represent actual performance or the quality of their safety management system or both? Is it based on a questionnaire or an audit in the field?

A common tactic is to assign additional costs in a bid evaluation to allow for the contractors and/or the company to provide additional supplies, safety inspectors, or training to close identified gaps. But an excellent safety culture is priceless. It takes time to develop, probably more time than the contract will last. The additional investments may provide incremental improvement in a focused area, but are unlikely to lead to the same performance as a contractor with an established safety culture. On the surface, a leader can use the tactic to explain the selection of an obviously less-safe contractor, but this veneer is easily scraped away by those looking for a Safety First mindset.

Think back to your last three major contract awards for services involving significant safety risk. Did you pick the safest contractor? Why not? How do you know? What drove the decision? How did people in your organization react?

Key Actions

- **Take bold action**
- **Insist on safety excellence from business partners**

Personal reflection and learning

- What have your leaders done to inspire you to make the workplace significantly safer?
- How satisfied are you with the pace of progress toward your safety vision? What would it take to increase the pace significantly? What is holding you back?

- How do you motivate employees to go beyond their normal responsibilities to improve safety? To innovate?
- How do you celebrate and recognize such behaviors and accomplishments?
- What obstacles might you face if you push for improvement too hard or too fast? How can you overcome them?
- How do you inspire safety excellence in your business partners? If you don't choose the safest contractors (with similar other capabilities), why not?
- Think of the last three changes you made to improve safety performance. How bold were they? If implemented well, what is the largest impact they could have had? How did they impact employee's behaviors? What kind of feedback did you receive?

Competent

Knowledgeable, well-informed, conversant.

Not unaware and disinterested.

Closely related to Curious.

A line leader is responsible for delivering all aspects of the business, but they can't be experts in everything. Many are not knowledgeable in all aspects of safety. That's understandable, but if Safety truly is First, or even Second or Third, they need to understand the most significant safety hazards in their work and how they are controlled. They also need to understand and participate in the systems supporting those controls, e.g., audits, risk assessments, and incident investigations.

How do you gain this knowledge amid your other duties? One way is to work in a Safety or HSE role as part of your career path. If you haven't done so, no worries, you'll find endless opportunities within your current role.

- Just ask—the safety staff, the frontline (what are you most worried about today?), peers (how do you manage this hazard?), your direct reports (what are the most significant hazards in your area of responsibility?).
- Participate in risk assessment exercises, e.g., HAZOP's (Hazard and Operability reviews), HAZID's (Hazard Identification reviews), Job Safety Analyses, etc. Don't just attend to show your support; participate and learn.
- Participate in Toolbox Talks.
- Review the Hazard Register, an inventory of the hazards at the site.
- Review incident trends and investigation results from the site.
- Understand the effectiveness of existing controls.
- Inquire about new hazards for proposed changes, new projects, and new activities.
- Review and discuss critical procedures and standards related to safety.
- Look outside your organization for better practices. Ask your peers, participate in industry associations, and seek to understand incidents outside the organization.

This sounds like a lot of work (and it is), but it can be integrated with the other aspects of your job. The learning should be continuous. Start with the most serious hazards and those unique to your site or business. Understand when and where the hazards occur, how they injure people, the controls used to mitigate the risk, how to fail safely, and common causes of associated incidents. This knowledge will help you develop improvement plans, ensure incident investigations reach the underlying causes, verify effectiveness of controls at the frontline, and ask relevant questions during any discussion.

A leader can also use their knowledge to test, influence, and send messages to leaders of the organizations they work with, including contractors. Can their leaders describe the most significant safety risk in their work and the corresponding controls?

While auditing, I commonly asked leaders what their top hazards were and how critical hazard controls were implemented on the site. Blank looks and referrals to the safety manager always piqued my interest.

***

Let's look at a project manager who truly viewed safety management as part of his job.

> Sophie was the HSE Manager on a large project to build a new blending and distribution plant, her third such project. She thought she understood how project managers worked, but Brett was different. He surprised her with his detailed questions about incidents, upcoming challenges, leading indicators, and many other aspects of safety. Previous project managers left that to her and the rest of the leadership team to handle.
>
> Like most of their projects, this one was a partnership with two other companies. Every month, a subset of the project leadership team met with the partners to share progress and discuss the key activities over the next few months. This was the first such project for one partner, so they brought people from multiple disciplines, sometimes even HSE. Their questions were sometimes detailed and numerous.
>
> Sophie prepared the HSE summary slides for the meeting, as her colleagues did for their disciplines. Project managers from previous projects had always asked her to attend to provide the HSE update. If she was unavailable, they reluctantly covered the material on the slides.
>
> Brett was different. Before the first meeting, Sophie asked, "Would

you like me to attend the meeting to provide the HSE update?"

Brett said, "You are welcome to attend if you are interested and available, but I can cover the HSE update. Thanks for providing the slides. They are right on target and consistent with what we have discussed the past few weeks."

Sophie raised her eyebrows. Her to-do list was long, so she reclaimed those two hours. "OK, great. Let me know if I need to follow-up on anything."

A month later, she asked the same question. Same response. "I got this." Another two hours freed up on her calendar. At first, she felt guilty about not contributing where she had in the past. She asked again every few months or when meetings with other stakeholders arose to make sure there wasn't a misunderstanding. But the response was always the same—unless he was absent due to leave or business travel.

When she asked his deputy the same question, the response was different. "Yes, please attend. You'll provide a better update. I'm not sure I could answer many of their questions." Granted, the deputy was preparing to cover many disciplines, but the contrast was stark.

As time went on, she realized Brett viewed this as part of his job. He believed line management owned safety. Her role was to support and challenge. Brett proved to her and the rest of the project team that he owned safety over and over, but he wanted to demonstrate it to their partners as well. In fact, he probably didn't even think about it much. It was part of his job as a project manager and a safety leader.

She let it go, used the free time to support the team in other ways, and was confident he would represent the project well. She knew he was prepared by the questions he asked her and his other discipline leads and the observations he made. He understood the risks, both the current ones and the future ones, and how the team would mitigate them. He understood why they failed occasionally and what they did to prevent recurrence. He was more than competent in safety.

How comfortable do you feel discussing safety in engagements with employees, business partners, regulators, and other stakeholders? When do you seek help? Why? What message would Brett be sending to his business partners if he addressed cost, schedule, and other project risks himself, but delegated the safety update to the HSE manager?

Do the same in town hall updates and safety events. Take the lead; don't delegate to the safety department. But know the material. The audience will figure out whether you know the material or are reading off a script. These are great opportunities to demonstrate your commitment and competence to the entire organization.

***

Demonstrating competence to the boss, peers, and business partners is important, but doing so at the frontline is critical. A leader who walks by an obvious, uncontrolled hazard without saying or doing anything sends a powerful message. "You tell us to follow the rules and work safely, but you don't take the time to understand the basics yourself." Or perhaps, "You saw it. You did nothing. You must think it is OK. Therefore, I can do the same."

Workers may form similar impressions when a leader asks questions that indicate a lack of understanding of what they do and how they are expected to manage safety.

This presents a dilemma for many leaders. If they fear missing something and sending a bad message, they may avoid visits to the frontline. But that is likely to show a lack of interest and care, an even worse message. I had mixed feelings during my site walks. I was exhilarated to see how our plans were being implemented in practice and by the opportunity to learn how I could help the workers be safer, but fearful I would miss something and send the unintended messages described above.

Leaders make mistakes too. They will miss some observations, but they need to demonstrate they know the basics of the safety of the work at their site. I was focused only on safety, and the burden was high. Most leaders are trying to observe multiple issues, like quality, progress, etc. There is no such thing as a quick walk out to the site to check out the progress on a painting job. *Every walk is a safety walk!*

A senior leader doesn't need to have all the answers. Most workers will understand they have many other aspects of the business to manage. They don't expect the vice president to weld, just as the vice president doesn't expect them to facilitate the next HAZOP. Ask specific questions about how their hazard controls are working, what obstacles they face, and how their work could be made safer. If you have a focus area for improvement, visit worksites involving that hazard or activity and focus questions on that area. Specific, relevant questions can demonstrate competence and curiosity (another key trait to be covered later) at the same time.

***

A proactive and competent safety department can assist in the improvement journey, but a leader shouldn't rely on them so much that he or she is viewed as distant, inactive, and uninformed. In addition, the safety department will never have as much influence on the rest of the organization as line leaders do. Some responsibilities of an effective safety department are to:

- Perform the administrative duties associated with incident management, audit planning and execution, leading indicator data collection and analysis, procedure management, government reporting, etc.
- Provide subject matter expertise on key safety risks in the business.
- Provide safety training.
- Seek out best practices from within and outside of the organization.
- Provide constructive feedback to leaders on the safety messaging of their words and actions.

We'll come back to the last bullet in Chapter 8. Some leaders liken it to being the conscience of the organization, a daunting responsibility.

The line leaders must lead safety in the organization, not the safety manager. Safety is part of the work, part of the business, not managed as a separate process by a separate group. The more you know about safety, the more you can help the safety department help you and the rest of the organization improve.

Key Actions

- **Understand the key hazards and controls in your business**

Personal reflection and learning

- What are the top three safety risks for the site or business? What controls are used to mitigate those risks? How effective are they in practice? What else can you do to reduce the risk? Why isn't it being done?
- Which hazards do you need to learn more about?
- How do you stay knowledgeable and current on the key safety risks at the site?
- How often do you refer inquiries to the Safety department? Which issues?
- What safety management systems do you understand well? Which ones

do you need to learn more about?

- How do you ensure you understand critical safety procedures, change proposals, and risk assessments?
- How much and how well do you address safety in engagements with your organization, business partners, and other stakeholders?

Humble

Modest, respectful, and self-accountable.

Not arrogant or self-centered

Related to Curious and Transparent (and Specific).

It's not about you, it's about them. Humble safety leaders strive to keep their workers safe out of care for the workers versus benefits to themselves or the company.

A humble leader:

- Focuses on those around them versus themself.
- Admits to their mistakes.
- Seeks input from others because they understand they don't know it all.
- Talks less, listens more.
- Recognizes the contributions of the organization, not the brilliance of their own ideas.

Self-centered leaders believe they have the best ideas and want to talk about them. They don't solicit the views of others, and when they receive comments anyway, they ignore or debate them. Their organization is full of people who understand the work in more detail. Humble leaders actively solicit their input, ask for solutions, and recognize their contributions.

***

Self-centered leaders rarely recognize the contributions and efforts of others. They tend to recognize only those who agree with them and follow their instructions.

> The project was complete, but the final year was a battle. A piece of critical equipment was delivered over six months late, so the construction and commissioning plans had to be redone after months of fine tuning the original plan. The plan for next week was developed the week before, including the safety planning. But the team rallied and prevented any serious injuries during that year of upheaval.
>
> The president of the company was in town for an international conference and asked to meet with the project leadership team to recognize their efforts, particularly how they kept everyone safe during difficult times. The invitation alone stirred their pride, and they anticipated the event.
>
> When the day arrived, the project leadership team gathered in a conference room around a large table with box lunches in front of them. Their host was ten minutes late. The boxes whispered, "Go ahead, open me up." Their stomachs said, "Go for it." But their minds told them, "It might be impolite to start without our host, despite his tardiness."
>
> Just as the project manager reached for his box, the company president entered the room. "Sorry I'm late. I was held up at the conference. A journalist from a trade group wanted some quotes about our next project. Go ahead, please start your lunch," he said as he opened his own.

After one bite, he continued. "It's been a busy week. I met with a few companies we might partner with on future projects. On Tuesday, I led a panel on project management excellence. I shared how the reorganization I initiated two years ago has led to ninety percent of our projects being completed on time and under budget. It's been a great week."

Everyone else around the table was having trouble swallowing their first few bites. He was supposed to be recognizing them, not himself. They didn't even have a chance to introduce themselves, to put names and faces with the amazing accomplishment he was taking credit for.

The project manager interrupted the monologue so his team could introduce themselves. The president ate quickly so he could continue when they were done. He even peeked at his smartphone. The project manager recognized the team at the end, but they already knew how he felt.

The president continued. "Yes indeed. Well done on an excellent recovery from a major setback. Further proof that our organization is now much better at project delivery."

His assistant cracked the door and indicated it was time to go. He whisked out of the room, and they all turned to their teammates mumbling, "What just happened? I thought we were being recognized."

Fortunately, the team was self-motivated. If not, they would not have recovered from the setback and delivered the project safely. They also recovered from the demotivating lunch quickly. They didn't need validation from the top of the organization. It would have been nice, but they didn't need it. However, their confidence in their leader was lower than ever.

But what if the team had not been so self-motivated? What if they had just suffered a setback and needed the additional empathy, confidence, and motivation to recover? What if they needed additional resources? How much would this narcissism motivate them?

Their success had little to do with the broader reorganization. This was a small event in the leader's calendar, but it had a large impact on his credibility. Unfortunately, when they shared the story with colleagues, they heard of similar experiences.

How much do you focus on your accomplishments versus those of the organization? How much credit do you take for the successes of the organization? How do you recognize the contributions and accomplishments of your organization?

***

Leaders are human. Those without humility may never admit it, but they make mistakes too. A humble leader recognizes their mistakes quickly, admits to them appropriately, and contributes to the recovery. Curiosity, another key trait under *Consistency*, will help a leader recognize their mistakes. They will check on the impacts of their decisions, actions, or initiatives versus assuming they had the intended effect. Admitting mistakes publicly also builds transparency. But don't do it too often, or people will wonder why you are in charge.

When recovering from a mistake, a leader should roll up their sleeves and contribute to the solution. This will show their humility is genuine; they are disappointed in themself, but still committed to moving the organization forward. Asking others to clean up the mess while they move on to their next priority may indicate a wavering of their commitment when the going gets tough. Do you want the rest of the organization to respond that way when the going gets tough?

To be honest with you, this aspect of humility is rarely observed, so it makes a strong impression when seen. Leaders will often defend their actions and espouse the benefits even when changing direction a year or two later. The bolder and more public the original idea, the less likely they are to admit a mistake. This lack of transparency and humility usually makes the organization reluctant to support future changes. A simple admission or apology won't win all the trust back, but is a step in the right direction.

When others make innocent mistakes, show empathy, coach them to show humility, and support them in learning from the mistake. Blame will discourage them and others to improve the status quo.

A leader doesn't need to wait for a mistake to demonstrate humility. They can simply admit they don't have a solution to a current challenge or aren't knowledgeable on a topic. They can seek advice from staff members or a consultant. Of course, the balance with competence is a fine line. Don't mistake curiosity for lack of knowledge.

***

I will introduce the idea of feedback here, but Chapter 8 will cover it in detail. Leaders should establish methods for understanding the safety messages they are sending with their actions and behaviors. This is how they Hear the Talk, as described in the Introduction. Seeking feedback demonstrates humility in a couple of ways. The very act of seeking feedback indicates the leader recognizes he or she can make a mistake and send unintended messages. When they discover a misstep, they can admit the mistake and work toward making it right.

Key actions

- **Focus on them, not you**
- **Admit your mistakes**
- **Check personal messaging**

Personal reflection and learning

- How much credit do you take when the organization accomplishes a significant milestone?
- In large engagements with employees, e.g., town hall meetings, how much do you talk about what is important to you versus what is important to the audience? How do you know what they are interested in?
- In small engagements, how much do you talk? How much do you listen? How do you encourage sharing by other participants?
- Think of a time you made a mistake on a safety-related matter. How did you respond? Did you make excuses? Did you publicly admit fault or apologize? How involved were you in the recovery?
- How do you respond when one of your direct reports makes a mistake on a safety-related matter?
- How else do you show humility?

Summary

In this chapter, I described traits that demonstrate the sincerity of a leader's commitment to safety. I covered sincerity first for a reason. Without it, you won't be able to send a positive and consistent safety message to the organization. Sincerity is about understanding your true motivations and developing your objectives so you can demonstrate your commitment consistently.

Below, I list some actions and behaviors people will look for to judge your sincerity. Their judgment is what matters. Their judgment will determine how effective your efforts to drive safety improvement will be. You don't have to display all the green flags, but the more you do, the stronger your credibility will become. Displaying a red flag occasionally will be OK, as long as it is not too significant, and you recover well. People in your organization realize you will make mistakes, just like they do.

Green Flags

- Get to know your employees as people, not as numbers, tools, or boxes on an organization chart.
- Describe your safety vision specifically. Identify specific plans to get there.
- When faced with dilemmas, act in complete alignment with your vision. Share your logic.
- Demonstrate care for injured persons and their family and colleagues after an incident.
- Learn the key safety risks in your business and their controls.
- Speak to safety yourself in town hall events and engagements with business partners.
- Engage with the frontline often. Focus on care and safety.
- Take bold steps to make material progress toward your lofty vision.
- Admit to mistakes and be part of the solution; seek advice from experts.
- Listen more, talk less; but ask insightful and relevant questions.

Red Flags

- Send vague messages about your safety expectations.
- Under-classify incidents or become too involved in the incident classification process.
- Blame incidents on worker error, or worse yet, stupidity.
- Walk by obvious safety hazards at the worksite.
- Be absent from the worksites.

- Undertake initiatives under the guise of safety when the true motivations are elsewhere.
- Focus on numbers, not people, e.g., after incidents.
- Boast on your successes.
- Continue to advocate a change and describe its success when the intended benefits are not achieved.
- Delegate everything safety-related to the safety department.
- Have no specific and material safety improvement plans.

Personal reflection and learning

- Which of the listed traits are most critical for demonstrating sincerity? Why? Which others are not listed?
- Which traits demonstrating sincerity do you view as your strengths? Your weaknesses?
- What feedback have you received from others regarding your sincerity? What did you do about it? How did you obtain the feedback? How representative is it?
- How many of the Green Flags do you exhibit? Which will you add to your repertoire next?
- How many of the Red Flags do you exhibit? How sure are you? What were the consequences? How will you reduce them in the future? Should you follow-up on any of them with the organization?

5

Consistency

Safety moments have become an obligatory opening to many meetings in organizations with significant safety risks in their work. The intent is good, engaging everyone in safety, but they are sometimes trite and unrelated to the real hazards in the workplace. In addition, they foster the concept of safety as a separate activity. *Safety is not a moment. It is part of the work.* It never ends. Safety must be a part of every activity, decision, or conversation where it is relevant. Hence, consistency is one of the two key pillars of safety leadership. A credible safety leader must repeatedly speak and act in accordance with their safety vision in different contexts.

This chapter describes the traits of a leader who demonstrates consistency. The traits of an inconsistent leader are generally the opposite and are not covered specifically. However, the stories describe a mix of leaders who did and did not demonstrate the trait.

Engaged (and visible)

Involved, present, and communicative.

Not distant or passive.

Closely related to Transparent (and Specific) and Curious.

An engaged and visible safety leader:

- Spends time with all levels of the organization, especially the frontline, learning about the work and discussing safety.
- Deliberately incorporates safety into all discussions and decisions where it is relevant.
- Recognizes outstanding contributions toward the safety vision.
- Personally speaks to safety in town hall meetings, recognition events, meetings with business partners, and other such engagements.
- Actively seeks feedback on how to make the workplace safer and how their messages are interpreted.

Safety moments, safety days, and safety meetings have become commonplace institutions intended to demonstrate ongoing commitment to safety improvement. In fact, some will associate their mere absence with a lack of commitment. But have they become a crutch to ensure safety gets its due attention? Are they necessary if safety is truly part of the work, whether at the frontline or in the C-suite? People can be hurt at any time during the day or night. Poor engineering decisions can be made at any time. I've always admired leaders who integrate safety into everything they do. To me, this was one of the clearest tells about a leader. Since their comments and inquiries are relevant to the discussion or work at hand, they are natural, not forced to fit within a designated placeholder. Recall the story in the Introduction about the morning meetings for the offshore oil and gas project. Of course, if safety is not relevant to the issue at hand, no need to force it. That can appear trite.

Some safety activities are natural for a leader to perform, such as progress reports and inspirational messages in town halls, handing out prizes at the annual safety recognition event, and kicking off a new contract. However, a leader shouldn't limit themself to these highlight reel activities, but rather participate in a wide variety of engagements to reinforce their commitment in front of their organization. Safety-focused engagements include:

- Participating in toolbox talks.
- Participating in hazard hunts or housekeeping rounds.
- Verifying the effectiveness of hazard controls at the frontline.
- Participating in risk assessments.
- Kicking off safety inductions or training classes.
- Participating in contractor safety reviews.
- Covering HSE in stakeholder engagements outside of the organization.
- Conducting safety readiness reviews for high-risk or infrequent activities.
- Attending team meetings to solicit safety-related feedback.

The best safety leaders go a step further than these discrete, safety-focused activities. Addressing safety in the course of normal business is even more impactful because it demonstrates that safety is an integral part of the business. In many cases, it's as simple as asking how a change or action will impact safety.

If a team member wants to reorganize their department:

- How will the safety responsibilities and accountabilities change?
- How will it affect supervision at the frontline?
- What safety competencies do the new leaders need to be effective?

If the site services manager wants to change contractors supplying lifting equipment and services:

- What is their past safety performance?
- Have their safety systems been audited? What were the results?
- How will they be inducted on the site's safety expectations?

If the plant manager is interviewing candidates to replace the maintenance manager:

- How does the candidate personally demonstrate their commitment to safety?
- What does No Harm mean to them?
- What changes have they made to improve safety at the frontline?
- How did they respond to a significant incident?

Other opportunities to integrate safety into business are:

- Staff meetings
- Change proposals
- Project review meetings
- SIMOPS (simultaneous operations) reviews
- Site visits
- Business partner selection and performance reviews
- Review of new business opportunities
- Employee performance reviews
- Budget allocations

Routine and meaningful engagements at the frontline are perhaps the most impactful. For a large organization, top leaders cannot engage everyone personally. However, word of the encounters will spread, and the reach will be larger than they think. A leader should go out with a learning mindset, not to police, shake hands, or smile for photos for the next newsletter. They should see and hear for themself how well safety controls are implemented at the frontline. If controls are absent or not effective, they can find out why on the spot. They can learn about working conditions and what they can do to help. And they will get to know their workers. However, this can be a double-edged sword. If not done well, it can do as much harm as good. Walking by obvious safety hazards displays a lack of interest or basic knowledge. Talking about yourself and your great safety programs versus listening to worker feedback displays self-centeredness and a lack of care for the individuals. Fortunately, worker engagement skills can be learned.

***

Let's look at an example of a leader who not only engaged in crucial discussion, but took immediate action.

> James, the project manager, and Nick, the HSE manager, were visiting a fabricator of specialized equipment at a remote location. Half of the facility's capacity was dedicated to their project. Nick visited frequently, but this was the first visit by James since the kickoff meeting a year ago. After a brief meeting with the site manager, Aaron, in the office trailer, they went on a tour of the worksites. Everything always sounded great in the office, but the site walk would provide the real update. Linda, company site lead, and her safety lead escorted the two visitors.
>
> James and Nick were appalled at the poor housekeeping. Tools were laying in areas with no ongoing work; some were rusted, indicating they had been there a while. Unused nuts and bolts, packing materials, and cable ties were everywhere. The visitors picked up the items that might trip someone or could fall to lower levels. They couldn't pick up everything, but they needed to send a strong message. They quickly filled their pockets and looked for garbage or recycling containers. When they couldn't find any, they piled their collections in a protected skid pan so they could pick up more items without dropping anything.
>
> A painter saw them and introduced himself as Mark. After introductions were complete, Mark said, "I see you found one of the causes of the poor housekeeping. I don't like working in areas

this messy, but it's hard to keep the worksites clean with no trash bins around. We've asked our supervisor several times, but nothing changes. Most of the crews have just given up."

"How often do you see Aaron out here?" asked James.

Mark grinned and shook his head. "Maybe once a quarter. And when he does show up, he's not very interested in talking to us. He's always on a mission."

James asked the safety lead to take a few photos of the piles of collected materials and told the painter, "I'll ask him to fix this when we get back to the office."

Linda whispered to the safety lead, who quickly disappeared.

James turned back to Mark. "Thanks for the feedback. What other obstacles to housekeeping do you face?"

"The containers will solve the problem for some crews, but some supervisors are so focused on progress that they push their crews right up to the siren at the end of shift. People are in such a rush to get out of here, they don't bother cleaning up. I wish they would just admit that cleaning up is part of the work."

"How can I help?" asked James.

"Maybe they could sound the siren twice. Once fifteen minutes before quitting time and then another at quitting time. And if the site manager came out once in a while to check that people were cleaning up before they left, that would help get the supervisors aligned."

"OK. Consider it done. I'll talk to Aaron about it before I leave. Linda, let me know what they decide to do and how well it is working. Mark, if it still isn't working in two weeks, let Linda know, and I'll give Aaron a followup call. I appreciate you giving such honest feedback to someone you don't even know."

"Thanks for listening. It hasn't worked very well with the leaders here, and I saw an opportunity with a fresh pair of ears." Mark paused and smiled. "I remember what you said about safety at the kickoff lunch. The poor housekeeping will get someone hurt eventually."

As soon as James returned to the office for the closeout meeting, Aaron was apologizing. The safety lead had snuck away to show him the photos and describe what was happening on the site visit. "James, I'm so sorry about the mess out there. I've seen the photos and asked the supervisors to stop work an hour early today to clean up the site."

"Thanks, Aaron. That's a good first step. What will you do to make sure it stays clean?" asked James. "And are you going to help the crews clean up this afternoon?"

Aaron hesitated a second too long; he knew his hand had been forced. After exchanging ideas and agreeing on corrective actions, James said to Aaron, "Please call me in a week to update me on the progress."

James' day was not done. Linda would receive some coaching over dinner. Hopefully, she had already received the gist of the message from participating in the site tour. Nick would be asked some tough questions on the ride to the airport the following morning.

Don't be fooled by the trivial nature of this example. Several people learned powerful lessons in a short time. James' request for updates ensured Aaron would spend more time at the worksite and would follow-up on his commitments. Linda, the company's site lead, discovered she had grown tolerant of poor housekeeping. She wondered what else she might be overlooking. Perhaps she had been sending mixed messages about schedule and safety. She would be more careful about her own words and actions. Mark remembered the safety commitments from the kickoff sessions and took the opportunity to check if James really meant it.

What would have happened if James and Nick had done nothing? Mark may not have spoken up. His skepticism about the company's commitment to safety would be confirmed. The tolerance for poor housekeeping at the site would grow. The mess would continue. Only an injury would change the course at that point.

What do you do when you see an isolated unsafe act or condition? How about when you see many? How do you follow up on your observations and requests? How do you encourage workers to share their true concerns?

***

In the previous example, the feedback came freely to the project manager. In many cases, leaders must be more creative and proactive in soliciting such feedback. A mix of specific and broad questions works well. Specific questions show that a leader recognizes the safety aspect of the issue and wants to learn more. Open-ended questions allow team members to bring forth issues the leader did not foresee. Examples include:

- What new hazards will be introduced by that change?
- How can we reduce the exposure to this hazard?
- What are you doing to address the increasing exposure to the hazard (or an increasing trend of incidents)?

- How will you prepare workers for the change?
- How do the frontline workers view the change?
- How well is our new PPE, safety procedure, etc. working? What feedback have you received from the frontline?
- What did we learn from the incident? What are we doing about it? How will we prevent it from occurring again?
- What's the most dangerous thing you did yesterday?
- What one change would improve your safety the most?

***

Leaders of organizations with a competent and well-resourced safety department may be tempted to rely on them to ask the questions listed above. And yes, they should, but leaders must also invest their own time and energy in safety. If safety is truly first, why wouldn't you spend a corresponding amount of time on it? When I visited sites for reviews and audits, I became concerned when a leader answered my questions with references to the safety department or their direct reports. Use the safety department for support, not as a replacement for line leadership.

***

Many of the examples presented in this section are routine tasks. But when the big stuff comes up, leaders need to be out front–the major incidents, improvement plan development, selection and support of focus areas, business partner selection, reacting to deteriorating performance, etc. If you already have strong credibility, people will look for you to set the tone and lead the way. If not, people will look for signs of your sincerity, or lack thereof. Credible leaders show up during a crisis and roll up their sleeves, not just ride in the victory parade.

Key Actions

- **Mix safety into everything**
- **Spend your own time on safety**

Personal reflection and learning

- Which methods of engagement on safety work best for you?
- How do you ensure you engage all parts of the organization on your safety vision?
- What safety-related activities do you participate in? How do you engage vs. observe?
- Think of examples of safety improvements arising out of normal business discussions versus a safety activity.
- How consistent are you in your safety messaging with different audiences, in different settings, in good times and bad?
- Which line leadership safety responsibilities do you rely on the safety department for? Why? How can you hold line management accountable instead?

Curious

Interested, inquisitive, and learning.

Not disinterested and assuming.

Closely related to Persistent, Competent, and Engaged (and Visible).

Excellent safety leaders continuously learn by:

- Actively seeking feedback on how to make the workplace safer.
- Checking the effectiveness of hazard controls at the frontline.

- Learning from incidents and successes.
- Engaging with peers in their own company and in the industry.
- Participating in safety studies and assessments.
- Understanding the rules and procedures workers are expected to follow.
- Asking about the safety implications of proposed changes, new business, and changing conditions.
- Challenging the quality of safety data.

Be curious in good times and in bad. Many leaders will search for answers and insights when serious incidents occur. What was the cause? Why did our systems and controls fail? How do we improve them? How do we improve hazard recognition? Hopefully not, "Whose fault was it?" Fewer leaders ask the same questions about less serious incidents, which may be just as valuable learning opportunities. The most curious leaders ask questions when things appear to be going well. With so much on their plate, leaders can be tempted to assume that good news or no news means they can spend their time elsewhere. Or they could show their curiosity by asking:

- Are we really that good or just lucky?
- Why are we doing so well?
- How good is the reporting/information?
- How are the controls for our top three hazards working?
- What new hazards may arise in our upcoming work?

Surely you've seen your share of scorecards with stoplights. Have you heard the expression *challenge the green and support the red*? Many leaders gloss over the greens and some shame the reds. But the reds need support, not shame. And are the greens really green? If the level of trust in the organization is low, supervisors or managers may be reluctant to share bad news. A green scorecard often keeps the second-guessers away. But what if a leader challenges the green? They may find that all is good and discover best practices others can learn from. Or they may find the information was bad and can remove obstacles or provide other help. Learning doesn't only come from incidents.

How does a leader find out if green is not red or yellow in disguise?

Trust, but verify. Some leaders interpret *trust* as "I can turn my back and assume it will happen." Others believe checking or verification is *policing*. However, if done well, verification shows care if the leader supports workers in removing obstacles.

One way for leaders to verify is to visit the worksites and work teams and see and hear for themselves. Excellent safety leaders can go a step further by implementing a more formal system for verification at the frontline. The concept is quite simple. For the most part, the controls needed to prevent injuries are known. If one measures how well the controls are implemented during the work, then they should know the relative likelihood of an injury.

Let's consider Dropped Objects. Examples of controls to check may be:

- Are tool lanyards secured at both ends?
- Are gaps or holes in grating covered with tarps, boards, or other barriers?
- Is an effective exclusion zone established below the work?
- Are exclusion zones free from unauthorized or unnecessary personnel?
- Are scaffolds fitted with toe boards and free from excessive gaps?
- Are loose items present at the worksite?

Unlike open-ended, worker engagement questions, these are simple Yes/No questions by design. The results of these inspections will show which controls are strong and which need improvement. Of course, the inspectors need to gather insights on *why* the deficiencies were present, so improvement actions can be targeted to the cause. The inspections should not be a tick the box exercise from afar, but include a discussion with the work crew.

Even with this type of verification system in place, excellent safety leaders will still check for themselves to verify the accuracy of the data, since safety data are notoriously biased toward the positive.

***

A mix of leading and lagging indicators is a powerful tool. When only measuring lagging indicators, the absence of incidents may cause a leader to ease up on continuous improvement. But were they that good, or just lucky for a few months? Leading indicators may provide the answer. Conversely, if the leading indicators are positive but the lagging indicators are not improving, the organization may have chosen the wrong leading indicators or the improvement plan is off target. Used together, leading and lagging indicators can guide the path forward.

In my opinion, the best leading indicators are close to the frontline and focus on what employees can control, not outcomes (as in injuries). The frontline verification data described above are generally an excellent choice. The data can be used to create simple charts of *Percent Implementation of Controls*. Other leading indicators, such as the number of workers trained on time and the number of awareness campaigns, are further removed from the work. These are easier to measure, but not as effective because they don't show how well the hazard controls are implemented. Just because someone was trained doesn't mean they are applying the controls every day. Data collection requires more effort, but you get what you pay for. Consider both the value and effort when choosing a leading indicator. Indicators with low value may give a false sense of security.

Make the data visible to the workforce in posters, emails, bulletin boards, and engagements. People want to see the results of their hard work. They may want to see early signs of success before going all in. They may feel peer pressure to do their fair share. The more proactive ones may recognize the plan isn't working as intended and suggest improvements to the plan.

***

Incident investigations are difficult and time-consuming. Team members are often taken away from their day jobs and eager to return before the make-up work becomes insurmountable. In some cases, the next incident needs resources for an investigation before the first is complete. Or the team members are not trained in root cause investigation methods.

A quality investigation should be initiated quickly, while the evidence and memories are fresh. Leaders should ensure the organization has a robust process with qualified investigation leaders and participants so they don't feel obligated to meddle in the details of the investigation. They should check in to see if the team needs any support or if early learnings need immediate attention, but should otherwise let them do their job. Getting too involved may foster suspicion that they are looking for someone to blame or protecting their reputation.

The investigation should identify the underlying cause(s). If the causes stop with the injured individual, the investigation didn't go far enough. This may lead the team to some of the leader's favorite systems or processes, or maybe even themself, but they need to understand it to fix it. If they could have done more to prevent the incident, they can admit it and contribute to the fix, thereby showing humility.

Improvement actions should be aimed at changing equipment, procedures, and standards, i.e., things that will last. Otherwise, the learning will fade over time. Reviewing an existing procedure in a safety meeting will lead to marginal, short-term improvement at best. Workers were already supposed to understand that information. Something else prevented them from applying it. That's what needs to be fixed.

Reviewing investigation results of more serious incidents with an incident review panel of peers, bosses, and subject matter experts can help validate that the underlying causes were identified and that the planned corrective actions will make work safer. This will also help spread the learnings to other parts of the organization.

***

The best safety leaders made me uncomfortable as a safety professional by asking questions I couldn't answer or proposing to implement programs I didn't suggest. I was the safety manager; I should have known the answer, or I should have made the suggestion myself. I was usually the one pushing the safety agenda. It was refreshing and intimidating when the tables were turned. Over time, I learned this was just an excellent safety leader doing their job. The temporary embarrassment was preferable to the frustration of constantly trying to sell proposals of my own.

> Manuel was presenting plans for a glove campaign at several retail gasoline station construction sites. Three workers at two sites had injured their hands in the previous two weeks, including a severe laceration that required surgery to repair. Maria, the project manager, sat on the edge of her chair, but gave Manuel an opportunity to finish his summary.
>
> When she sensed a break, she said, "Manuel, excellent plan. I like the idea of trying three different models of impact-resistant gloves to gather feedback from the workers. If they don't use them, the glove program won't do any good. What about other types of gloves?"
>
> Manuel was ready. "We asked the contractor to prepare a glove matrix, which will tell workers what type of glove to use for each hazard. That will be ready in two weeks. But since our recent injuries were related to impact, we are working on that type of glove first."

"Sounds good, Manuel. Thanks for expanding the scope to all hazards. I agree that it's best to focus on one thing at a time. When is the glove trial going to occur?"

"Next Wednesday at the Creekside Avenue site."

"Please send me the details. I'd like to visit the site, try on the gloves myself, and ask the workers how they feel about the different types."

Manuel raised his eyebrows. "You don't need to go. Rachel (the safety lead) and I will be there and will gather the information. I can bring samples to your office."

"I'm glad you two will be there. I want to understand the pros and cons of the different types so we can make the best decision. The workers may have ideas we didn't even think of. I look forward to seeing you there."

Of course, Maria could not afford the time to do this for every challenge they faced. She trusted Manuel, but she also knew hand injuries were hard to prevent and many other hand safety campaigns had failed. She wanted to ensure the program would prevent hand injuries, not just show she was responding to the injuries. She was curious. She wanted to receive input from multiple sources. Not only would she learn a few things, but in the future, her project team would be more likely to explore all options in detail, seek her input, and share their challenges. The contractor's leaders would be more likely to support the improvement program. The workers would appreciate her interest in their input and their safety. She may even receive feedback on other hazards while there. So many benefits from such a small investment.

Key Actions

- **Verify controls at the frontline**
- **Use leading and lagging indicators**
- **Learn from incidents, don't blame**
- **Spend your own time on safety**

Personal reflection and learning

- How do you obtain information on how well your safety controls and systems are working? How good is the information? How do you know?
- How do you respond to no news or good news on safety performance? How do you *challenge the green and support the red*?
- How do you specifically increase your knowledge of managing and leading safety?
- Who do you get most of your safety information from? How representative are they of the organization?
- What safety performance metrics do you use? How well do they represent reality? How do you know? Which are leading and lagging indicators?
- Think back to your last three incident investigations. Did they get to the underlying causes? Have the improvement actions made the site safer?
- How else do you show curiosity?

Action-oriented

Responsive, proactive, and thorough.

Not passive, reactive, and assuming.

Closely related to Persistent, Focused (and Organized), and Inspirational (and Bold).

Safety leaders take effective actions to back up their words and head toward their vision:

- Develop and resource improvement plans with specific, measurable actions.
- Follow-up on actions from improvement plans, incident investigations, audits, reviews, etc.
- Verify completed actions are having the desired effect.
- Act on feedback from staff and observations in the workplace.
- Spend their own time on safety vs. delegating it all to the safety department.

Lack of action, weak actions, and poorly implemented actions undo thousands of inspiring words. As discussed above, bold actions can have an enormous impact on safety performance and a leader's credibility, but they should be complimented with a steady stream of smaller actions to improve continuously. Most organizations can only handle a few bold changes at once.

The actions should be specific and aligned with the agreed targets and focus areas. Actions like:

- Identify and implement opportunities to reduce working at height exposure during the next turnaround.
- Institute a diverse, frontline safety committee to provide monthly feedback to the plant manager on safety improvement opportunities.
- Formally train three staff in root cause analysis. Each person will conduct at least two root cause incident investigations to further develop their skills.

versus actions like:

- Reduce strain injuries by 50%.

- Update safety induction training.
- Automate trending of safety observation data.

The latter actions may indeed be worthwhile, but they are too general or may be considered business as usual.

Proactive, forward-looking actions should form the bulk of the improvement plan, if possible. However, actions can originate from multiple sources:

- Annual business planning
- HSE Management reviews
- Incident investigations and trends
- Audit results
- Leading and lagging indicator data analysis
- Worker engagement programs

Excellent safety leaders will only support proposed actions if they are convinced the changes will make a positive and material difference. Otherwise, they distract from those that will make a difference, provide a false sense of security, and expose a leader's lack of commitment to their safety vision.

A leader should make it clear which actions they are committing the organization to, who has responsibility, and when they will be completed. Make them visible to the organization, just like the safety performance data. People will become suspicious and assume inaction if they do not receive information to the contrary. If an action is dropped or modified after further analysis, share the reasoning with the organization.

Set action owners up for success. Are you providing enough resources to meet your expectations? Are you coaching when the first signs of trouble arise? Would you even know if trouble is arising? Those are the leader's responsibilities. Resources include money, additional people, time, leadership support, etc. Build the estimated resources into action plans.

Once the responsible person develops more detailed plans, ask if they need additional resources or guidance. One common obstacle to progress is assigning part-time team members without carving out space in their existing agendas. What can you take off of their plates so they can dedicate enough attention to the action plan?

***

Good intent and energy abound when an undesired event occurs and actions are assigned. However, the enthusiasm wanes over time as people return to the grind of everyday work and actions from other events accumulate. Excellent safety leaders follow-up on the status until closure and verify the closure was completed as intended. Others may hope the actions are forgotten and don't impact their work.

Since actions come from many directions, compile them in one place to make tracking and communication more efficient. Some call it a safety (or HSE) action register. A simple spreadsheet will do in most cases, but more sophisticated tools are available.

Review progress in leadership team meetings. Hold leaders of the responsible parties accountable by directing inquiries their way. Be prepared to support those who struggle.

Verify closure of a sample of actions yourself or through your leadership team. By verify, I mean to check that the document was revised and issued, impacted parties understand the changes, new equipment is in stock and available, resources are in place to support the change over time, etc. Workers will look for and count on these things to improve their safety.

Finally, discuss the status of key actions in workforce engagements, big and small. This demonstrates the commitment to improve and provides the opportunity for feedback on how effective implementation is.

Here is a story of a leader who did this well.

One of the biggest challenges in managing safety at facilities that handle hazardous materials is balancing personal safety and process safety. The controls and required skills are different. In an operating facility, the need to manage both well is obvious. However, in projects, the emphasis changes over time. Process safety tends to dominate the early stages of a project, since most process safety controls are established in design. Good safety leaders, however, will simultaneously plan the construction phase of the project, where personal safety tends to dominate the attention of leaders. Toward the end of a project, the construction activity subsides, and the focus returns to process safety, as the project and operations teams verify the facility is safe to start up.

One key performance indicator (KPI) used to keep focus on process safety during the construction phase of a project is the number of overdue safety action items, most of which come from technical studies during design, e.g., HAZID's, HAZOP's, and other risk assessments. On one of my projects, our performance was outstanding, i.e., only one or two overdue actions every few months, out of hundreds. My peers were jealous or suspicious when comparing this to the several dozen overdue actions they had every month.

When we compared notes, the difference was clear and simple. My project manager had drawn a line in the sand early in the project that zero overdue items were acceptable. We had automated tracking, monthly reporting, and early notifications of nearly overdue actions. In addition, the project manager had to approve extensions of the high risk actions. He was willing to sacrifice his valuable time to review and follow-up on these actions. Needless to say, that act alone meant he received few such requests. His leadership team knew he would follow-up if something was close to being late. That was the difference from the other projects, which left it to the HSE and the engineering teams. Their leadership team accepted dozens of overdue items every month, and that became the norm. You get what you tolerate.

How do you ensure action plans are completed, and completed well? Can you give an example where you assumed an action was completed, but later found it was not? How did that make you feel? How do you verify closure of actions? How well do you balance your attention to different aspects of safety (and health)?

***

Leaders should take on some actions themselves. Delegating everything safety-related is a clear signal that Safety is Not First. Also scrutinize how much effort the organization invests in safety improvement compared to improvement in other aspects of the business. Many people will assume that the most time and effort will go to the first priority.

The following demonstrates how easily an organization's priorities can be gleaned from observing the actions at the worksite.

Companies which manufacture large steel equipment such as ships, buildings, tanks, etc. have high safety risk, but also must compete on cost, schedule, and quality. I visited one company which took pride in the world-class quality of their products. Their safety performance was below average, with several severe injuries every year.

Their commitment to quality was hard to miss. During quarterly performance reviews, they presented detailed data on weld quality by type of weld, product, month, subcontractor, etc. They even tracked weld reject rates by welder. Welders with poor performance were observed, coached, and retrained. For safety, I saw the typical lagging indicators, a few generic leading indicators, and highlights from the past quarter. The contrast was stark. I suggested that if they managed safety with the same rigor as welding, they could be world class in safety as well. The safety manager nodded, but the rest looked puzzled.

I was escorted for a site tour on a cold, brisk morning the next day. Dropped objects were one of the most significant safety risks at the site. As I approached a storage tank under construction, I noticed heavy-duty orange tarps covering a scaffold, which formed a tunnel around the perimeter. I pointed toward the tarp with my stiff fingers and tried to smile with my numb lips. "That's a good way to prevent dropped objects from falling beyond the work area. How do you decide where to use the tarps?"

My escort replied with frosty breath, "Oh, those tarps aren't for dropped objects. They're used to control the climate for the welders. They can weld under a wider range of weather conditions. The quality is better, and delivery is faster."

My enthusiasm and optimism were short-lived. I doubted I would see the same tarps on a beautiful spring day.

The company had proven they could be the best when they focused their attention. Welding was a critical success factor in the business. They measured it, they found the root causes of under-performance, and they invested in their people and equipment. They were action-oriented. Sound familiar? Imagine how many injuries they could have prevented if they had directed similar effort toward improving safe work practices?

In what part of the business does your organization perform best? Why? If not safety, how can you apply what you learned to becoming better in safety? Where does most of your focus on improvement go? How about the rest of the organization?

Key actions

- **Take bold action**
- **Provide resources**
- **Spend your own time on safety**
- **Follow-up on action plans**

Personal reflection and learning

- In which parts of your business do you focus most of your improvement efforts? If not safety, why not?
- How material are your safety improvement efforts? Will they really move the needle?
- How do you identify safety improvement actions? What is the most important one at this time?
- How well balanced is your safety action plan between proactive and reactive actions?
- What input do you gather from the organization before committing to actions?
- How do you organize and monitor actions being taken to improve safety performance? What do you do when they go off-track or are not having the intended effect?
- How do you verify actions are completed and have the desired effect?

Focused (and organized)

Purposeful, coordinated, and systematic.

Not aimless, unsure, and unplanned.

Closely related to Action-oriented and Persistent.

A focused and organized leader:

- Focuses on the big stuff, i.e., high-risk hazards, biggest opportunities for improvement, emerging hazards, etc.
- Sets clear priorities for improvement, with input from the organization, i.e., makes the tough choices.
- Ensures the actions they commit the organization to are manageable and will lead to the desired improvement.
- Uses a safety management system to bolster the effectiveness of hazard controls at the frontline.
- Is not distracted by the day-to-day noise.

Hazards are diverse and everywhere, and the controls and systems to manage them seem overwhelming. It's tempting for a leader to say they want it all and want it now. But that will just overwhelm the organization, and may indicate the leader does not understand the problem well enough to focus their efforts.

McChesney, Covey, and Huling state in their book, *The 4 Disciplines of Execution*, "the more you try to do, the less you actually accomplish." If too many changes are attempted at once, none may be successful. In my previous book, *Don't Let It Fall: Stopped Dropped Objects, Save Lives*, I showed leaders how to focus their improvement plans on dropped object prevention. While that in itself is a broad and diverse problem, overall safety improvement is a massive challenge.

No matter how a leader defines *Zero*, they must get the big stuff right, especially for the hazards that can kill people. Unfortunately, there are lots of them, so a leader faces the challenge of selecting one (or a few) to focus their improvement efforts on. They can't ignore the rest, but the organization can probably only drive material and sustainable improvement in one or two at a time. Hopefully, the organization already has strong programs for the others. A leader may want to focus on those actually causing injuries or serious near misses. Or perhaps those that are new or on the rise due to a project, turnaround, or new business. The personal knowledge they developed on the hazards at the site will be valuable here, but they have an organization to provide additional input.

Selecting a focus area can be challenging, but is the easy part. Turning it into meaningful actions and sticking with it are the hard parts, the parts that save lives. Some of the types of actions a leader can take are listed below:

- Develop a specific action plan aimed at the focus area.
- Develop and monitor key performance indicators specific to the focus area.
- Reinforce their focus and the case for action in multiple engagements with employees and business partners.
- Celebrate success along the way, while driving continued focus (unfortunately, success is rarely permanent in the safety business—only when the hazard is eliminated).
- Adjust for changing risk profiles.
- Focus their attention on the selected area during their site visits.

This trait overlaps with Action-oriented and Persistent. See those sections for discussion of action planning and follow-up.

***

Here is an example of how a leader successfully transitioned the safety focus as the hazard profile changed in a project.

The project was coming to a close. Construction of the new process unit was complete. The commissioning team was ready to implement the plans they had developed over the past year. The operations team was trained on their new procedures. For the past two years, the project team had focused on preventing falls from height, one of the most common causes of fatalities in construction work. They had upgraded their fall protection equipment, reduced work at height where feasible, improved their scaffold inspections, reviewed their contractors' fall protection plans, and measured the effectiveness of their frontline controls. They used the verification data to create a leading indicator (percent effectiveness of fall protection controls), which was reviewed every month in the project leadership team meetings.

Now, the permanent work platforms and grating had been installed, and most of the scaffolds had been removed. Very few open holes should be present during commissioning. The potential to fall still existed, but the likelihood was much lower, and the controls were strong after two years of focus. New hazards would emerge. The project team discussed where to focus next. They would soon be pressure testing many of the piping systems with water or nitrogen. The pressures were high enough to create jets that could pierce the skin or turn loosely connected fittings into high-speed projectiles. They chose this hazard as their focus for the next few months.

They worked with their contractor to add valuable details to their procedures, for instance, how large the exclusion zones should be for different pressure ranges. They ensured workers were fully trained and that everyone else was aware of the hazard and the controls. They participated in the SIMOPS (Simultaneous Operations) meetings. And they generated another list of controls to verify at the frontline and track as leading indicators. When the workers saw the leadership team give this hazard the same attention as they had to fall protection, they knew they had to get it right. The controls were nearly always in place when checked, and there were no injuries or serious near misses associated with that hazard.

Now, on to start-up. What would be the focus area for this phase?

How many focus areas do you have? In safety? In other parts of the business? How many can you effectively manage at once? How often do they change? How do you know when to shift focus?

***

A Safety Management System (SMS, or HSE MS, if it includes Health and Environment) is a collection of principles, work processes, and tools working together to manage safety systematically. In other words, it helps to organize all aspects of your safety program. Haphazard and/or contradictory procedures, forms, and initiatives frustrate workers.

When effectively using a HSE/Safety Management System, a leader should have access to a plethora of information on their existing safety program: incidents, leading indicators, audit results, safety observation trends, Safety Committee feedback, etc. A leader can gather this information and analyze it with their leadership team. This is called an HSE Management Review, typically an annual event where leaders review a wide variety of information relevant to HSE Performance. Potential changes to safety exposure in the business plan and the regulatory climate can also be considered.

Good questions for discussion include:

- What are the largest gaps between your performance and your vision?
- What are the common causes of underperformance?
- What tactics have been effective in previous improvement efforts?
- What feedback are you receiving from the workforce?
- How well is the leadership team aligned on the vision and the pace of change?
- What areas need immediate attention?
- What actions are needed to build the capacity of the organization to continually improve?
- What are the top two or three areas the organization should focus its safety improvement efforts on during the next year?
- How will you measure progress on implementation of those priority actions?

The answers to these questions will be invaluable in developing your safety improvement plan for the next year.

Key actions

- **Focus on the big stuff**
- **Follow-up on action plans**

Personal reflection and learning

- How do you select your focus areas for safety? How do you involve others?
- What is your focus area for safety improvement currently? How does it stand out? How many do you have?
- What action plans do you have in place to support it?
- How do you balance focus areas related to safety and those in other parts of the business?
- Do you use a Safety Management System? How well does it work? How could it be improved?

Persistent

Determined, steady, and unrelenting.

Not wavering, content, and assuming.

Closely related to Action-Oriented and Curious.

A persistent safety leader:

- Develops robust improvement plans and follows up on implementation.
- Stays the course in the face of setbacks, tweaking their plans as needed, based on learning.
- Recognizes outstanding contributions and good performance along the journey.
- Maintains chronic unease about safety performance in themself and the organization.
- Continuously seeks opportunities to improve.

Persistence is an element of several of the preceding traits: Curious, Action-Oriented, and Focused (and Organized). I considered removing this trait from the list, but kept it because of one unique aspect: longevity. That alone is worth a short discussion here.

The journey to Zero Harm is long; most will argue it never ends. For every two steps forward, the organization may take one step back. Therefore, persistence is critical to continually moving toward the vision. Accomplishing the vision is not a project that will end with a clear victory.

Demonstrate commitment through words and actions in good times and in bad. Periods of good, even great, performance may tempt a leader to back off. They may want to move on to the next business priority or give the organization a break from the unrelenting pressure. While recognition of progress is needed along the way, persist with the drive for continuous improvement. Unfortunately, the second law of thermodynamics seems to apply here, i.e., left unattended, work will tend toward chaos. It takes considerable effort just to maintain the status quo, much less move forward.

Chronic unease means never being content with safety, always wondering what could go wrong next, what to do about it, and how else to improve. Balancing recognition and chronic unease is tricky. Chronic unease can be exhausting, but continuous improvement will not happen without it. Some leaders are reluctant to celebrate successes for fear of jinxing future performance, but it is necessary to show people that their tireless efforts are working and worthwhile, and to prepare them for even tougher work ahead.

A persistent leader doesn't let a few incidents distract them from their improvement journey and plans. Unless something relevant has changed, they stick with the overall plan, learn from those incidents, and adjust as needed. Changing direction after every injury may signal several weaknesses: the leader doesn't know how to improve, they aren't confident in their plan, or they are chasing short-term metrics on their scorecard.

***

Let's look at an example of a leader sticking with their plan to the very end of a project, despite many temptations to change focus.

In the section on *Inspirational (and Bold)*, I described a project's journey to reduce its high-risk dropped objects, including verification of the effectiveness of dropped object controls at the frontline. It was a long journey, but the project manager, Janice, persisted until the last day of the project, three years later. They had no significant injuries and reduced potential serious dropped objects by over 80%.

Janice's actions were simple individually. The difficult part was the consistent emphasis over the rest of the project, while the team faced many other challenges, even crises. She insisted on monthly meetings on dropped object prevention with her leadership team. The work sites held their review meetings weekly. Those meetings were closer to the work and more important, but she wanted her leadership team to demonstrate their commitment and help the large number of sites around the world learn from each other. The HSE Manager gathered the verification data from the work sites, and they reviewed it during the meetings.

She asked probing questions about any trends. "I see Site A has improved their use of tool lanyards. How did they do so? Now the effectiveness of their exclusion zones is falling. What are they doing about it? What can we do to help?" If no new trends were visible, "What is the next challenge? What will we do about it?"

She focused on dropped objects more than any other single topic while visiting project sites. She asked about the challenges the site faced and what was working well, then went to the workplace to see and hear for herself.

She challenged her leadership team to replicate effective work practices and controls at all the project sites. She influenced other projects and other operating units at the plant to implement their learnings. She recognized the team for their individual and collective success in front of each other and senior leadership of the company.

She led like this for three years. The HSE manager had tried to implement changes like this in other projects and operations, enough to realize the change would have faded away without this support.

Many project managers would have switched their focus after meeting their objective during the first year. After all, projects like these are filled with hazards that can seriously injure people. But she knew the hazard remained, new contractors would come on board, and the types of dropped objects might change. She even stayed the course during a flurry of minor hand injuries that garnered the attention of her boss. Could you have stuck with it for so long?

Stick with improvement plans until the intended goal is accomplished and the elevated performance is sustainable. Don't be tempted to move on to the next challenge at the first signs of success; real change takes time. The organization will be less likely to support future initiatives if their leader abandons the previous ones before they are successful or if performance quickly deteriorates.

How do you know when to switch your attention from one safety improvement program to another? What information do you use to decide? Think of an example where an improvement plan failed because its support waned too quickly?

Key actions

- **Stick to your improvement plans; adjust as needed**
- **Follow-up on action plans**

Personal reflection and learning

- How do you respond when seeing success immediately after a safety improvement action?
- How do you respond to challenges to move on to other priorities?

- How far into the future does your safety improvement plan extend? One month, one year, five years?
- How do you balance the strain of continuous improvement with celebrations along the way?
- How else do you show persistence?

Appreciative

Grateful, indebted, and generous.

Not thankless, inconsiderate, and selfish.

Closely related to Caring and Humble.

Appreciative leaders:

- Recognize not only excellent safety performance, but contributions that move the organization toward their safety vision.
- Consider safety behaviors as significant factors in awarding promotions and filling empty roles.
- Reward business partners contributing to their safety vision.
- Give credit where it is due.
- Balance recognition with persistence for continuous improvement.

Recognition and reward for safety performance evoke spirited debate in leadership teams, control rooms, and even social media. I don't claim to be an expert on the subject, but in my experience, celebration of progress and success is necessary when deserved. The path to *Zero* is arduous and never really ends, leading some to believe what they do is never good enough. Well-deserved recognition can break the vicious cycle and re-energize people for the next step in the journey.

Individuals, teams, and business partners should be recognized for excellent safety performance, completion of significant improvement actions, and other efforts to move toward the safety vision. Recognize both results and efforts. Safety incidents can occur somewhat randomly over shorter time periods, and better performance may lag improvement efforts. Connect current and past improvement efforts to better performance when possible. Recognition that employees' hard work resulted in better performance can boost confidence in the future improvement plans.

Some leaders are reluctant to recognize excellent performance for fear of jinxing future performance. Such superstition precludes any recognition at all, since the improvement journey never ends.

Safety rewards are often criticized for fostering under-reporting or under-classification of incidents. While this is a legitimate concern, a leader can mitigate the risk by balancing recognition between leading and lagging indicators and reacting appropriately when incidents occur. In fact, consistent and accurate reporting of incidents, especially near misses, may be a behavior to reward in itself.

Behaviors and actions worthy of recognition include:

- Applying relevant and material learnings from outside the organization.
- Extensive mentoring of less experienced staff.
- Proposing solutions for long hidden hazards or new trends in incidents.
- Designing tools or work aids to reduce exposure or eliminate hazards.
- Leading and delivering significant improvement efforts.
- Providing honest and constructive upward feedback to leaders.
- Intervening under difficult circumstances.
- Completing deep and insightful incident investigations.
- Improving systems to make them user-friendly with no sacrifice in effectiveness.

Be careful not to recognize other business achievements reached using behaviors inconsistent with the safety vision. This encourages others to take shortcuts on safety and causes people to wonder whether you noticed the safety implications of the shortcut. Examples include reducing cost while the maintenance backlog of safety critical equipment grows or completing a turnaround early by asking fatigued craftspersons to work extended shifts for weeks.

A leader can also reward excellent safety contributors with promotions or broadening opportunities, as discussed later in the section on *Fair (but Firm)*.

*****

Don't forget about your business partners, including subcontractors. Perhaps the simplest and most powerful recognition is to continue using their services or products and increase their scope, if possible. How often are contracts rebid for the sake of achieving a lower cost? You will see an example of this in the section on *Fragility* in the next chapter.

Another reward mechanism for contracts is the use of incentives. Some leaders are reluctant to use incentives for safety, thinking it should be a given, but don't hesitate to provide incentives for meeting cost or schedule targets. When they are used, they are often dwarfed by those given for cost and schedule. What does this tell the contractor's leadership about your priorities? Will it lead them to meet schedule milestones or cost targets at the expense of safety? If you are a welder receiving zero safety incentives, what message do you get when your supervisor asks about progress five times a day?

*****

Finally, an example demonstrating that big checks are not needed for effective recognition. In this case, the inclusion of the workers' families made it special.

The crane cabs were empty, the forklifts were parked, and yellow tape marked routes for the visitors to take on their tours of the oil and gas platform that would soon be installed offshore. Two days of progress were being sacrificed, one to clean up the site in preparation for hundreds of visitors and one to show off their work to their families. But it was a small price to pay. The platform would be towed offshore earlier than promised, and their safety performance was the best of any such project done by the company.

Those who could brave the elevator ride and walk across the gangway hovering over one hundred feet above the water had their photo taken on the helipad where their mother would land every two weeks, touched the twelve-inch diameter pipes installed by their husband, and bounced on the beds their father would sleep on for half of the year. Of course, the most popular stop was the galley, where nearly everyone partook in the cookies and ice cream. Clearing that area for the next group was a challenge for the tour group leaders.

Back onshore, under the large white tent, the project team members enjoyed their BBQ lunches as their spouses and children met some of the bosses circulating the room and listened to others at the podium. Family members laughed as they heard stories not shared by their spouse and grinned with pride when the name of their loved one was mentioned by one of the speakers.

For those on the project team, the end was near. They would move on to the next project. For those who would operate the platform, their turn was just beginning. The next time many of them would be onboard, they would be surrounded by deep blue water, swaying in the long swells. Everyone would remember this day for the rest of their careers.

Key actions

- **Recognize excellent performance and behaviors**

Personal reflection and learning

- How do you recognize employees for contributing to safety improvements? What are the most effective methods you have used?
- Do you reward performance or behaviors and contributions more? Why?
- How fairly is recognition spread throughout the organization?
- Can you describe a situation where you gave a significant award to a person who took shortcuts in safety? Or deliberately did not give recognition when shortcuts were taken?
- How do you recognize contractors with outstanding safety cultures and performance?
- How do you evaluate the safety capability of a contractor before selection?

Fair (but firm)

Principled, unbiased, strong, and consistent.

Not prejudiced, wavering, or disproportionate.

Closely related to Caring.

A leader demonstrates fairness by:

- Focusing on learning versus blame when incidents occur.

- Recognizing contributions from all parts of the organization.
- When appropriate, applying discipline consistently and proportionate to the behavior.
- Including safety behaviors and performance as material factors in conducting performance reviews, filling jobs, and granting promotions.

A couple of related concepts are important here. The first is treating everyone fairly with respect to their personal accountabilities. The second is treating everyone fairly with respect to each other. These are multi-dimensional concepts, but I will focus on the safety aspects here.

One of the most visible and controversial aspects of fairness is how *rule-breakers* or those who *cause* incidents are treated. I won't get into all the differences between one-strike/you're out, just culture, and other schemes for managing discipline, but an organization should have a well-defined system that is fair by design and in implementation. An undefined system is susceptible to inconsistent and unfair implementation.

A common flaw in implementation is to blame individuals who are closest to an incident or who break a rule, without understanding the circumstances leading to the event. In most cases, systems failures and/or poor leadership lead to the behaviors for which the individuals are blamed. Assigning blame and disciplining individuals often indicate a leader's refusal to take accountability for his or her role in the event. It can lead to under-reporting of incidents or rule-breaking, especially when related to behaviors where individual discipline is commonly enforced. Unfortunately, these are generally associated with the serious hazards you want to manage the best.

***

Be sensitive to applying discipline fairly between different groups of people: staff vs. hourly, supervisor/manager vs. individual contributor, and high-flier vs. average performer. While such discipline is generally kept behind closed doors, the general outcomes are readily apparent and can be discussed widely among workers. Not only does unfair treatment diminish the credibility of the decision-makers, but it can also create rifts between different parts of the organization.

The cooling water treatment system was acting up again. "We probably need to tweak the chemicals," thought the operator, Jeremy. He dialed his supervisor, who was responsible for all utilities at the plant.

"Hey Esther. I think it's time to call in Sammy from the chemical supplier. We've been struggling with cooling water quality all week. We may need to cut back flow soon, and that means cutting back production. I don't know what else to do. Every time Sammy comes out, he tweaks a few of the chemical dosages, and the water gets cleaner."

"That might be a problem, Jeremy, but I'll get you help."

"What do you mean, a problem?"

"The last time Sammy was here, he failed his drug and alcohol test and was banned from the site. We'll probably get somebody else this time."

"I haven't been impressed with the people they send when Sammy is on vacation. Can we get an exception?"

"Let me check. I'll get back to you." Esther hung up and walked to the operations manager's office.

"Hey Barry. I need your help. We need someone from the chemical supplier to come out and troubleshoot our cooling water problems. Sammy is the best they have, but he failed his drug and alcohol test last time he was here and was banned from the site. Can you get a waiver for us?"

"I don't know Esther. That's our policy."

Esther wasn't giving up so easily. "Sorry to cause trouble, but we may only be a day or two from impacting production. I remember one of our own maintenance technicians received a warning when he failed his test. Maybe Sammy should get a second chance?"

Barry looked down and mumbled. "OK, I'll send an email to Contracts and Security approving his site visit. I hope he can solve the problem quickly."

The inconsistency in discipline sent mixed messages. Esther remembered the exception made for the maintenance technician, which led to an exception for Sammy. The site policy was clear for failed drug and alcohol tests onsite; one strike, you're out, whether or not the manager agreed with it. But business priorities prevailed, production in this case. The next exception would be even easier to grant. The site policy was almost meaningless.

What message was being sent to the chemical supplier? To the other contractors onsite? Probably that the company was playing favorites or production rules safety, both of which hurt the credibility of the decision-maker.

How consistently do you enforce the discipline policy on safety issues? What other factors influence the decisions? How fair are they?

∗∗∗

Other leaders and staff in the organization should include their contributions to safety targets and improvement plans in performance agreements or personal goals. Performance reviews should include discussion of progress on these contributions, but also how the leader is supporting the safety vision overall. How are they demonstrating the traits outlined in this book? What messages are they sending with their words? Review observations of their behaviors and what messages they may be sending. Ask them to give feedback on your safety leadership as well.

Include safety performance as a key factor in their performance appraisal. If safety truly is first, shouldn't it be the most influential criteria? A leader can mitigate concerns about under-reporting of incidents by basing the evaluation mostly on fulfillment of agreed commitments and actions and leading indicator performance.

Safety should be an integral part of resourcing decisions as well. To a large extent, a leader's credibility in safety is determined by that of their direct reports. Imagine the impacts of promoting a maintenance team lead who reduced costs while building a large backlog of safety-related repairs as the new maintenance manager. The credibility of the new maintenance manager will remain low, and the leader's will suffer as well. If Safety was truly First, why was a poor safety leader appointed to a position where safety is a big part of the job? Since so many factors go into such decisions, only the egregious ones will be noticed. But since they are egregious, they can impact credibility a lot.

Key actions

- **Empower leaders and hold them accountable**
- **Learn from incidents, don't blame**
- **Recognize excellent performance and behaviors**

Personal reflection and learning

- How do you deal with behaviors clearly not aligned with the safety vision?
- How well is your safety-related discipline policy documented and understood? How fairly is it applied across the organization and with business partners?
- How much is safety leadership considered when filling leadership roles? How do you evaluate the safety leadership performance of candidates?
- How much weight is placed on safety contributions and performance in employee performance reviews? What feedback is provided to the employee?
- How deeply are broader causes of incidents evaluated before placing blame on individuals and applying consequences? How are leaders' contributions to the incident considered? Disciplined?
- How do you balance fairness and firmness?

Summary

Green Flags

- Routinely engaging with employees and teams at all levels of the organization on safety expectations.
- Recognizing excellent safety performance and contributions toward the safety vision.
- Rewarding contractors and other business partners with excellent safety performance.
- Insisting on thorough incident investigations to discover underlying causes.
- Producing and maintaining a focused and cohesive safety improvement plan.
- Following up on action items from improvement plans, incident investigations, and other feedback mechanisms until they are complete and

effective.

- Effectively using a Safety Management System to improve safety at the frontline.
- Maintaining a consistent focus on one or two critical risks.
- Verifying implementation of hazard controls at the frontline.
- Learning from successes, failures, and other people and organizations.
- When appropriate, applying discipline consistently and fairly.
- Asking insightful questions about safety; participating in safety studies and assessments.
- Considering safety contributions and performance in filling jobs and awarding promotions.
- Taking on some safety actions themselves.

Red Flags

- Trying to improve everything at once.
- Changing focus areas and improvement plans after every incident.
- Engaging in safety only in planned, safety-related activities.
- Taking actions that contradict your claimed safety vision, especially when the stakes are high.
- Blaming individuals for causing incidents when leadership or systems are the underlying cause.
- Filling key positions managing large safety exposures with leaders with low safety credibility.
- Selecting a contractor with low safety capability over one with a proven safety record, when they both can deliver the desired services or goods.
- Applying discipline unfairly.
- Delegating everything safety-related to the safety department.
- Taking personal credit for organizational successes.
- Allowing improvement actions to go uncompleted.

Personal reflection and learning

- Which traits listed are the most critical for demonstrating consistency? Why? Which others are not listed?
- Which traits demonstrating consistency do you view as your strengths? Your weaknesses?
- What feedback have you received from others regarding your consistency? What did you do about it? How did you obtain the feedback? How representative is it?
- How many of the Green Flags do you exhibit? Which ones will you add to your repertoire next?
- How many of the Red Flags do you exhibit? How sure are you? What were the consequences? How will you reduce them in the future? Should you follow-up on any with the organization?

6

Credibility

According to *dictionary.com*, credibility is "the quality of being believable or worthy of trust." The traits described in the previous two chapters build credibility in different ways. As you have seen, many of them overlap and complement each other.

- A **caring** leader is motivated by the interests of people in the organization, not just their own interests or those of the company.
- A **transparent** leader is open and honest. They are **specific** in their expectations. They tell it like it is.
- An **inspirational** leader makes excellence a believable target people can rally toward. A **bold** leader places their reputation on the line for the organization to achieve excellence.
- A **competent** leader possesses the knowledge to be believed and trusted.
- A **humble** leader admits when they make mistakes or don't know something. Oddly enough, this makes them more human and ordinary, and therefore, more believable.
- An **engaged and visible** leader gives people plenty of opportunities to judge the conviction of their mission through their words and actions.
- A **curious** leader is always learning how to make people better and the work safer.
- An **action-oriented** leader supports their words with consistent actions

that make a positive and material difference.

- A **focused and organized** leader knows what needs to be done to improve and ensures that it happens. They focus on the big stuff.
- A **persistent** leader doesn't give up when the going gets tough, showing how much they believe in their mission.
- An **appreciative** leader knows they cannot reach the vision alone and recognizes those who make significant contributions to the journey. They understand their drive for excellence can wear people down.
- A **fair but firm** leader expects excellence, not perfection, from people in the organization, and can be trusted to learn and show empathy when mistakes are made.

Most leaders will be expected to demonstrate all these traits to some degree. Fortunately, most people in the organization will understand a leader is stronger in some areas and weaker in others, just like themselves. However, leaders must be strong in a few critical traits to build strong credibility. The relative importance of the traits will vary somewhat by individual, business, and organization. In my opinion, the following are most critical to building credibility:

- Caring (number one by far)
- Transparent (and specific)
- Engaged (and visible)
- Fair (but firm)
- Action-oriented

Why is care number one? If a leader cares, they will do what it takes to protect people from harm. Most of the other traits come naturally. If they care more about wealth, prestige, or power, they will be trying to fake their safety commitment every day. I say *trying,* because very few can really fake it. You can't just tell yourself to care; you have to really feel it. People can tell.

In large part, the strength of a leader's safety credibility determines how the organization responds to their action plans, requests for change, and attempts at inspiration. Those with credibility have the opportunity to increase their credibility further and improve safety performance dramatically. They are often given the benefit of the doubt, since their efforts have been sincere and effective in the past.

Those without credibility face stagnation, at best. They will be met with apathy, skepticism, or distrust. They will find it difficult to motivate the organization to follow their lead. Their actions and behaviors will be scrutinized even more closely as people in the organization try to figure out their true intent. These leaders may not get the full support needed to make their efforts effective.

It may not seem fair, but that is the power of credibility.

This chapter describes characteristics and outcomes of credibility, and the lack thereof.

Characteristics and outcomes of credibility

Fragility

Credibility is difficult to earn, but easy to lose. People need to hear the words and see the actions they expect in multiple contexts before they will trust a leader. Two types of behaviors can destroy credibility. The quickest way is to take one action or make one statement when the stakes are high, which contradicts their professed vision, especially if it threatens people or is viewed as dishonest or evasive. Credibility is lost immediately. Apologies or explanations are rarely effective. The stakes were high, which is when people expect a leader's true colors to show. Credibility is like a snowball rolling downhill, gathering size, until it hits a tree trunk and shatters into hundreds of pieces.

The other way takes longer, but the result can be similar. A leader loses credibility by regularly showing less significant inconsistencies with their professed vision. Some of their actions are supportive, others are not. People will wonder if their leader is having trouble faking it or has changed their priorities. They no longer know what their leader stands for and will look for more evidence before restoring their trust. The snowball rolls downhill, sloughing off chunks as it goes, never really growing in size.

***

Let's analyze an example of the first type, an immediate loss of credibility.

The new module was nearly complete. The transport vessel would arrive in two days, and the module would be lifted and set on board the following day. After being secured to the vessel, it would begin its long journey to a chemical plant halfway around the world. The module had to leave by next week if it was to arrive in time to be connected to the plant during the next turnaround. Otherwise, it may sit idle for a year until the next turnaround.

For the most part, construction had gone well. It appeared they would meet the schedule. The costs were on target, but they had struggled with safety. In fact, those struggles jeopardized the schedule. Six months ago, a scaffold pole fell from a height of forty feet into an unprotected area, landing just ten feet from a scaffolder. The next day, a ten-pound valve fell twenty feet while being hoisted to the third deck for installation. It landed on the foot of a passerby, sending him to the hospital for a week. He still could not work, but he had been inches or seconds from losing his life. The project manager stopped work for an entire week to purchase more equipment, provide additional training, clean up the site, and make sure all the leaders were aligned on priorities.

That week nearly cost them an on-time delivery.

Penny, the safety lead, rushed into Grant's office. Grant was the site manager for the client overseeing construction of the module. He and Penny would move back home soon. She placed one end of a short scaffold plank on the floor and leaned it against his desk.

"Another hazard hunt for dropped objects, Penny?"

"I wish that were the case, but this is the real deal. It just fell from the upper deck and landed six feet away from one of my safety inspectors, Larry. It could have killed him if he had been walking a bit faster. We suspect the scaffolders left it behind last night. They have been rushing to remove the scaffolds before the module is loaded out."

"Wow. Is he OK?" Grant then thought, *I'll be glad when that module is on its way.*

"He's OK, but was quite shaken. He's seen a lot of injuries over the years, so he is recovering quickly. He's organizing a hazard hunt for tomorrow morning. I'll enter the incident in the reporting system. It should be ready for your approval before you leave. By the way, I plan to classify it as a high potential near miss."

Grant raised his eyebrows. "OK. I'll be looking for it."

The next morning, Penny walked with Grant to the quayside to participate in the toolbox talks. She asked, "Did you have any questions about the incident notification I sent last night?"

"No questions, but I didn't approve it yet. I'm going to wait until the module is on its way."

"But Grant, these things are due within twenty-four hours. The incident might be a sign of people rushing to meet the shipping date."

"After the last incident, the project manager stopped work for a week. If he does that again, we'll miss our departure date. Just ask your inspectors to keep a close eye on things for a few more days."

"They're already working fourteen-hour days. Are you sure you want to do this?" Penny paused for Grant to react, but he didn't. "Think about the message you're sending to the rest of the team here. And I don't think our bosses back at the plant will appreciate the late notification."

"I'll take that chance. Let's manage it at the site." Grant veered off toward one of the work teams starting their toolbox talk.

Five days later, Grant approved the incident notification, and it was automatically emailed to the project manager and the HSE manager back at the plant. *They probably won't even notice since the module left on time*, Grant thought as he hit the Submit button.

His phone rang an hour later. "Grant, this is Stephen. What is this high potential near miss report all about? The notice says the incident occurred almost a week ago..."

Grant was badly mistaken, and Stephen, the project manager, was disappointed in himself as much as Grant. Somehow, Stephen had created the impression that the schedule was more important than safety. He needed to find another way to demonstrate this was not the case to the rest of the project team.

Stephanie and Grant had reported previous incidents as prescribed in the procedure. Fearing another work stoppage by the project manager, Grant made an exception this last time. After all, it was just a late incident report. What difference would it make? But it wasn't the late report that cost Grant his credibility as a safety leader; it was the motivation behind it. He had demonstrated that schedule was more important than safety. To those in the know, he would be remembered for one of his last acts on the project versus his efforts to improve safety at the site.

How do extreme business conditions affect your decision-making on safety? How would people in your organization react to this type of behavior? Have you ever hidden bad news to avoid scrutiny, delays, or extra work? Why? What messages were you sending others?

Don't take hard-earned credibility for granted. Credibility is powerful, but fragile...and very hard to earn back.

Trust (and Respect)

A credible safety leader will be trusted to take the right actions to improve safety performance for the right reasons, i.e., so people go home every night unharmed. People will seek to understand their leader's proposed changes and do their part to contribute. If the connection to their own safety is not clear, they may even give their leader the benefit of the doubt. After all, their leader had not let them down yet. They will observe the impacts of the changes and how their leader responds to challenges in implementation. Since the leader has the trust and support of the organization, they have a greater chance of successfully implementing material changes. If the changes make the workplace safer, the trust will grow, enabling even bolder changes.

✳✳✳

The leader in this example did not have ill intent, but the stakes were as high as they could get. He lost the trust of some and the respect of others.

The mammoth drilling rig was nearly back to work. It was a floating structure connected to the seabed with massive mooring lines to keep her in position while drilling. She had returned to the well location after spending a month in a shipyard for unexpected repairs. The mooring lines had been reinstalled, so they were nearly back to where they were before they left for repairs.

It had been a rough summer. Before leaving for repairs, a hurricane and a tropical storm had threatened their operations. The company evacuated all personnel for the hurricane, costing them a week of delay. In hindsight, the evacuation was not necessary, as the storm veered to the east. That was moot, as the decision had been made several days before then to prepare for departure. For the second storm, they secured the well, but did not evacuate. The storm did not strengthen as expected and continued along its forecasted path, well east of their location. Another false alarm.

Now a third storm threatened their progress. It was not yet a hurricane, and forecast to stay south of their position.

The rig manager onshore, Carlos, was checking the forecasts every three hours as they were issued. He awaited the word of potential evacuation from the corporate emergency response team (ERT). The company had five rigs in the region and managed hurricane response from a central team due to the massive resources required to evacuate personnel from the many facilities operating in the region. Carlos finally received word from the ERT that they should stay in position and not proceed with running the riser (the large pipe separating the drilling string and fluids from the sea). They would not evacuate this time. This storm appeared to be another false alarm.

Carlos forwarded the official email from the ERT to the offshore installation manager, Terry, and called him. "Hi Terry, I presume you saw the email from the ERT. It looks like they believe the storm will not threaten the rig, just like the previous two. I wanted to check if you needed anything."

"Yeah, I saw it. To tell you the truth, I'm very surprised. The forecast has been changing frequently. Who knows where that storm will go? About a quarter of our crew lives near the forecast landfall location. Their minds are already on the helicopter. I will not be a popular guy."

"Well, you know how it goes; these decisions are made by the ERT. I really don't have a lot of input."

"OK, we'll prepare the rig. It should be pretty easy since we haven't started drilling yet. We'll secure everything on the deck and get ready to hunker down. I better get on it. Talk to you later."

"Good luck."

Two days later, the storm strengthened quickly to a Category 1 hurricane and passed a lot closer to their location than forecast. Terry had been through a few storms like this early in his career. He was concerned as the living quarters rocked back and forth for hours. Chairs slid across the room, binders fell off the bookcases, and many of the crew were sick. Some were not only sea sick, but also terrified, especially the young ones who had experienced nothing like this.

Terry had heard enough. He dialed Carlos. "Hey Carlos. It's me again."

"Hi Terry. I appreciate the updates during the storm. It sounds like it was a little rougher than expected, but we should be able to run the riser shortly."

That was not what Terry wanted to hear, but it opened the door for the venting he had held back for twelve hours. "Carlos, I can't believe you left us out here. I was afraid the mooring lines were going to snap. Most of the new hands were scared to death. Several want to be sent home for good, others want to go check on their homes. I have a lot of work to do before we run the riser. People are so furious and distracted right now, I'm afraid someone will get hurt. Why did you leave us out here like this?"

"I'm sorry, Terry. I didn't realize the storm caused so much trouble. The ERT made the call. I guess they might have become complacent after the false alarms earlier this year. I'm under a lot of pressure to make up for those lost days in the shipyard."

Again, not what Terry wanted to hear. He needed to return to calming the crew and arranging a helicopter for those who had seen enough of the lack of concern for their safety and their families back home. Terry and Carlos had always told the crew they had a duty to stop work if they thought it was unsafe. But that wasn't an option this time. They were at the mercy of the managers onshore, people they rarely, if ever, saw face to face. The crew felt betrayed.

Being betrayed in a life and death situation will leave an impression a leader may not be able to change for years, if ever. This one decision, or lack of action, undid years of outstanding safety leadership by Carlos and Terry. Yes, the decision was complex. Most people knew that, but to them, it came down to cost and production versus safety. No explanations or excuses would be good enough for something with such severe potential consequences. Carlos had lost the trust of Terry and the crew offshore.

Back onshore, Carlos reflected on his discussion with Terry. Maybe he should have done more. He could have discussed it more with Terry to gauge the feelings of the crew. He could have engaged with the ERT, perhaps reminding them of the emotions offshore. Next time, he would do better. But for now, he needed to go offshore and face the wrath of the crew. More on that later.

How well do you identify critical safety decisions and insert yourself? How much does complacency or the lack of previous incidents affect your decisions?

Support

With trust and respect, a leader will get the support from employees he or she needs to implement changes effectively. Employees demonstrate their support by:

- Doing their part to implement the change.
- Being patient for the benefits to be seen.
- Providing feedback on obstacles to implementation.
- Encouraging reluctant colleagues to provide their support.

The best idea on the planet will probably go nowhere if that support is not provided. Of course, many leadership positions come with power. If a leader attempts to use it to force support, they risk diminishing their credibility even further by not being transparent, engaged, or fair. They may get some people to go through the motions, but true commitment to make the change work will be lacking.

Approachability

Credibility makes a leader more approachable. Employees are more likely to provide solicited and unsolicited feedback. They believe their leader truly cares for them and wants to make their work safer. By helping their leader, they help themselves. If their leader is sincere and fair, they will not fear retribution for providing feedback that a recent decision may have unintended consequences or may be impossible to implement.

As with most of the impacts of credibility, genuine feedback has the power to build credibility further. The feedback allows a leader to better understand the impacts of their actions and decisions and how others interpret them. The best leaders will actively seek that feedback, as employees will be unlikely to volunteer as much feedback as a leader needs. More on that in Chapter 8.

Leaders with low credibility will be flying blind. Many will not be interested in feedback, but if they manage to pry some out, the feedback may be guarded and unhelpful. They must find other ways to determine what effects their words and actions are having on the workforce.

Once again, the credible safety leader's job appears to be much easier, but this was earned by years of hard work.

***

Acting on feedback opens the door for even more.

> The plant manager walked into the control room. A few heads turned his way, but for the most part, everyone continued what they were doing. However, one of the field operators took his hard hat off and headed his way. "Hi Andrew. It must be Friday."

Andrew said, "Come on, Earl. Am I that predictable? I need to mix it up to keep you on your toes."

Earl chuckled. "We're here every day, and you're welcome anytime. Hey, I wanted to thank you for repaving the parking lot. Now we don't have to scrub the floor mats in our trucks every time it rains. Plus, it was just a matter of time before someone twisted their ankle."

The parking lot repaving was recommended in the monthly meeting Andrew had with the Safety and Culture Advisory Committee two months ago. The committee consisted of a representative group of frontline workers, including some of the major contractors. They gathered input on potential improvements in safety and general work culture at the site and had a monthly meeting with Andrew to discuss recommendations.

"You're welcome. I use the parking lot closest to the administration building, so I wasn't aware how much the other parking lot had deteriorated. I should have noticed how dirty the entrance road was getting. I'm sure it's irritating to end a long day by getting your feet wet and your truck filthy."

"You got that right!" said an eavesdropper across the room. "Much appreciated."

"Earl, you were headed outside when I came in. Where are you going?"

"Over to the truck loading rack. The drivers are telling us the loading pump is making weird noises."

"Do you mind if I walk with you?"

"Not at all. Let's go."

How do you think the operators in the control room would have reacted if Andrew did not show up every week? Or if he did not act on the recommendations of the committee? Or if he only showed up after an incident? Andrew didn't always act upon the recommendations of the committee, but he always listened and explained the basis for his decision. With that track record, he often received authentic and valuable feedback during his site walks.

How do employees respond when you approach them unexpectedly? How openly do they respond to your questions? Which groups respond better than others? Why? How well do you follow up on recommendations from the frontline?

Innovation

No matter how good a leader is, they can only move the organization so far with their own ideas. Excellent leaders inspire others in the organization to develop and implement their own ideas to improve the business, including safety. In some cases, they can just point in the right direction, and the other leaders in the organization (both formal and informal) will find ways to get there. Such leaders can get away with less visible demonstrations of the traits listed here. For instance, other leaders may develop improvement plans and engage with their teams. They will monitor implementation and adjust as needed. The top leader must still be engaged, supply resources, and recognize the efforts to encourage such proactive efforts in the future. Being humble in these situations is critical, but easy.

Characteristics and outcomes of a lack of credibility

Apathy

Apathy may result when leaders want to do the right thing, but are not engaged or inspiring and do not take actions that will make a material difference. The case for employees to change their behaviors is not compelling. Their leader has not made a strong connection between the proposed changes and their personal safety. Requests to make small changes may be supported, but larger changes may not. Safety advocates and highly motivated individuals may drive some change, but their reach will be limited.

Skepticism

Skepticism is one step beyond apathy. Chances are, the leader has been inconsistent with the messaging from their actions. Some have been effective and aligned with the safety vision, but others have not. Employees feel they need to scrutinize each action or request before they give their support. If the connection to the vision is not clear or the implementation plan is poor, they may not support the proposal or change their behaviors.

Distrust

The most extreme impact is distrust, which often leads to disregard. This is usually driven by a lack of sincerity. Employees have seen too much evidence that their leader is not genuinely committed to their safety. Their leader's actions are intended to tick the box on their scorecard or move them up the career ladder. Most employees won't even seek to understand the changes they propose. At best, they will go through the motions to avoid repercussions, which often accompany this style of leadership.

Some employees may use this to their advantage, particularly those aligned with the leader's true priorities. They will use the leader's misguided words or actions to promote their own agenda of cost or production over safety.

Recovering from this position is difficult. These leaders are not provided the opportunity to make incremental improvements to safety to rebuild trust. Much bolder demonstrations are necessary. More on that in Chapter 9.

Stagnation

A common theme of the lack of credibility is the reluctance of employees to support changes intended to improve safety. This results in ineffective changes, which can lead to a dearth of change proposals. This downward spiral results in stagnation of performance, or worse. If the stagnation persists, it can become part of the culture. New employees will pick up the apathy, skepticism, or distrust without even witnessing the actions that caused those feelings. Another difficult position from which to recover.

Summary

- Credibility is hard-earned and precious. Don't take it for granted.
- Credible safety leaders are given the permission and opportunity to further build their credibility because employees trust their actions and contribute to their success.
- Those without credibility face an uphill battle just to convince employees their proposed changes have good intentions, much less will work.

Personal reflection and learning

- How credible are you as a safety leader? How do you know?
- What was the last significant change you attempted to improve safety performance? How well did your leadership team accept it? How well did the frontline support it? How effective was implementation?
- What feedback do you receive from your leadership team and frontline workers on your proposed changes?

- If you have not yet achieved credibility as a safety leader, where on the apathy to distrust spectrum are your employees? What have you done to overcome the lack of support?

7

Actions Speak Louder Than Words

Many of the examples in the last few chapters illustrate actions, decisions, or behaviors that are not consistent with a Safety First vision. That was deliberate, since they provide more opportunity for introspection and analysis. Unfortunately, many of the supportive behaviors go unnoticed. The inconsistent behaviors (large or small) and the bold supportive behaviors are the ones that stick out. It doesn't seem fair, but that's the way it is.

However, the purpose of this book is not how to be a bad safety leader, but how to be an excellent safety leader. It's one thing to understand what traits are important for safety leaders, but traits are exhibited through actions and behaviors. In the first draft of this book, I tried to separate the traits from the actions and behaviors, but as you have discovered through reading Chapters 4 and 5, it is difficult to describe a trait without giving examples of actions and behaviors. The key actions were listed at the end of each section describing the traits, with some appearing multiple times.

In this chapter, I list the key actions in one place in a structured way. The material is repetitive, but I think the consolidated list will be helpful.

To help you understand how they fit together, I have organized them by:

- Establish yourself
- Prove yourself
- Improve yourself

Actions in the *Establish* category demonstrate a leader knows what their safety vision is and how to move toward it. Leaders *Prove* themselves by acting in support of the vision. They show they can drive change that improves safety performance and is sustainable. As you might expect, this is the largest category. Leaders solicit, receive, and analyze feedback from many sources to *Improve* as safety leaders. Some actions could fit into multiple categories, but I encourage you not to dwell on that.

<u>Establish yourself</u>

- Show care for people, not numbers
- Be specific about your vision and expectations
- Learn about the key hazards and controls
- Focus on the big stuff
- Focus on them, not you

<u>Prove yourself</u>

- Mix safety into everything
- Spend your own time on safety
- Follow the rules yourself
- Provide resources
- Take bold action
- Follow-up on actions and commitments
- Share the true motivations behind key decisions, especially dilemmas
- Insist on safety excellence from business partners
- Recognize excellent performance and behaviors
- Hold leaders accountable

<u>Improve yourself</u>

- Verify controls at the frontline
- Use leading and lagging indicators
- Learn from incidents, don't blame
- Admit your mistakes
- Check personal messaging

A nice, round twenty actions are listed above. Going back to *The Talk*, most of these actions fit into the *Walk the Talk* category. The last one is clearly about how to *Hear the Talk*. Of course, I could go on for pages. However, one of the actions is to focus on the big stuff, so I have tried to keep the list manageable.

I can almost hear questions leaping off the pages. "Twenty is still too many. Which are the most important?" The cop-out answer is they are all important. I said earlier that managing safety was the hardest part of a leader's job. I meant it. The next less useful answer is that the most important ones will differ from leader to leader. That is true, but in my opinion and experience, the top five are:

- Show care for people, not numbers (once again, by a landslide)
- Mix safety into everything
- Take bold action
- Verify controls at the frontline
- Learn from incidents, don't blame

"If I am a new leader, where do I start?" Much of your initial focus should be on establishing yourself. That is why I organized the actions this way. But before you even get to that, make sure you understand what your vision is, specifically. I covered this in detail in Chapter 2. *Know the Talk.* That being said, opportunities and challenges will arise while you are still establishing yourself. React to incidents in a positive manner. Take bold actions if the opportunities arise, as they don't occur often for some. If you are still unsure of your judgment, seek even more advice from your organization. Needless to say, the faster you establish your foundation for safety leadership, the more likely your initial actions will be consistent and start to build your credibility. A new leader will be under extra scrutiny, but may also be given some benefit of doubt as they adjust to their new role. However, large errors in judgment will still damage credibility.

The twenty actions listed here are quite general. In the discussion of each trait in Chapters 4 and 5, I gave examples of more specific actions. In Chapter 10, I will give more specific examples for different types of leadership roles, from the CEO to the construction foreman to a thought leader.

Summary

- Expected traits of an excellent safety leader are demonstrated by actions and behaviors, much more so than words.
- Actions can be categorized as *Establish yourself, Prove yourself, and Improve yourself.*
- Above all else, show care for people. From there, many of the others will come naturally.

Personal reflection and learning

- What actions have you found effective in demonstrating your commitment to excellence in safety? Why do you think they were so effective?
- What actions have you taken in the past that sent unintended negative messages about your safety commitment? What did you do about it?
- Select five actions described in this chapter. How do you personally demonstrate them? What else do you plan to do based on reading this book?
- What are your three weakest action areas? What specific actions can you take to reinforce your safety commitment?

- I took a shot at listing the top five actions. Which do you think are the most important? There is no right or wrong answer.

8

Using Your Mirrors

This chapter is about *Hearing the Talk*, briefly described in the Introduction.

Since a leader's intended messaging is clear in their own mind, they are sometimes blind to how their words and actions are interpreted by others. One role of a safety professional, especially a Safety or HSE Manager, is to hold up a mirror for leaders in the line organization so they can see the impacts of their words and actions as others might. This is one of the most difficult aspects of a safety professional's job. It must be done respectfully and selectively and mixed with praise (when appropriate). In fact, line leaders should point the mirror the other way occasionally.

Safety professionals are not the only ones who can and should provide such feedback. In fact, the view of the safety professional will sometimes be biased, perhaps looking for potential messages that are not there. Feedback from different sources and of different types provides a more complete view of the impacts of a leader's words and actions. This chapter describes a variety of methods for obtaining this invaluable feedback. Most leaders won't need all the methods, but should consider their options carefully.

Seeking feedback on messaging is a great way to demonstrate humility. By inquiring, a leader is admitting they can make mistakes like everyone else,

150

but are eager to learn from them.

Leaders who have not earned the trust of the organization face a dilemma. Most of these options rely on open and honest feedback to be effective. If a leader has created a fear of retaliation or shown defensiveness, the most valuable feedback will not be shared. The leader may receive feedback on trivial messages viewed as safe to share or, even worse, feedback that all is good when it is not. If a leader is still building trust, they should remain open, keep the discussions constructive, express genuine appreciation for the feedback, and act on it.

If the only feedback a leader receives is positive, they are not getting it all, no matter how credible they are. It may be time to try another approach or seek feedback from a part of the organization they have overlooked. And visit the worksites. Even if the verbal feedback is sparse, observations of work practices, effectiveness of hazard controls, and working conditions will be valuable. The more a leader is out there (and showing genuine interest), the more likely the verbal feedback will come.

***

Here is an interesting story where a leader held the mirror up to me so I could do the same for their bosses.

> A new process skid needed for a plant expansion project was being built ten thousand miles away. The local project team received few visitors from the project leadership team at the plant. But they had conference calls aplenty, usually in the evening local time to accommodate the normal working hours at the plant. The local team was worn down after non-stop action at the construction yard during the day. Generally, the construction manager, Timothy, was the only one dialing in from the plant.

As the deadline for delivery approached, schedule, setbacks, workarounds, resourcing, and deliveries dominated the discussion. The local company representative, Anthony, was concerned about the increasing number of near misses that were seconds or inches from injuring someone. He filed the incident reports promptly and tried to discuss them during the calls, but at best, Timothy asked if they had re-trained the workers on the proper PPE and controls.

Then, a welder fell through an open hole in the grating during the night shift. He was wearing personal fall protection, but not the correct type for the location. He struck a valve before his fall was arrested. Fortunately, he only received a painful contusion, but it could have been much worse. Anthony was scared and needed help. He needed to get the attention of the project manager, Leslie. He was impressed by her safety messaging, insightful questions, and detailed observations during her two visits. A peer of his had told him of an intervention she made at another site when the investigation of a serious near miss did not get to the root cause. Timothy wasn't going to help, so he tried another avenue.

"Hi Anthony. I read your email when I got home. It sounded like you needed to talk right away. What's up?" said Susan, the project HSE manager. Susan had visited the site with Leslie on those two occasions and another time with Timothy. She had the ear of Leslie whenever she was concerned.

"Thanks for the call. I didn't mean to disturb your evening, but I wanted to give you a heads-up about an incident last night and ask for your help. Can you come out here in the next couple of days?"

"First thing first. Was anyone hurt? What happened?"

"A welder fell last night and badly bruised his hip, but he could have been hurt much worse. This is the first serious near miss in a while, but we've had other less serious ones as the delivery deadline approaches. I really think you should get out here." Anthony described the incident in detail to satisfy Susan's questions.

Susan returned to Anthony's request about a visit. "I can leave here tomorrow if need be, but let me talk to your site safety lead and get his perspective."

"OK, but hear what I am saying…you really need to get out here!"

"I understand. I'll check if I can book a flight tonight, and I'll send you an itinerary so you can arrange a pickup at the airport."

***

Susan arrived at the site right before the daily progress review with the company's site team. She felt the tension before she even sat down. While she was traveling, a pipe fitter had dropped his wrench while tightening nuts on a flange. It landed on the ground fifty feet below, five feet from hitting an electrical inspector.

The mechanical team lead said, "We need more resources from the contractor. They are working their crews fifteen hours a day, seven days a week."

The safety lead said, "We also need better SIMOPS (simultaneous operations) planning. It's going to get tougher during commissioning, but the contractor keeps saying it will be easy for such a small project."

And on it went. If Susan didn't know from Anthony's call why she was here, she did now.

Anthony walked up to her after the meeting. "Let's go for a walk."

As they walked to the module, she said, "I think I see why you were so insistent. I imagine my observations will confirm that."

Anthony winked. He would have smiled, but he was too scared and too tired.

Susan witnessed many people working alone, tools and materials scattered on the grating, scaffolds missing toeboards, and a dozen other hazardous conditions and behaviors. Too many to intervene on individually, though they tried their best.

***

Back at the plant, after a sleepless night on the plane, Susan headed straight for Timothy's office. "How did your visit go? I didn't know you were going until you were on the plane."

"To be honest with you, not good, but I needed to go. The site team is frustrated. The contractor is behind, but they're not adding resources and are letting their safety behaviors slip."

"These projects always seem like they're behind schedule at this point. The site team should be used to that by now. Anthony and his team just need to watch them more closely."

The discussion went on, but Susan knew she would get nowhere with Timothy.

It was late afternoon before she could meet with the project manager.

Despite being the end of a long day for both of them, Leslie welcomed her with a smile. "What's up? How was your trip? I'll let you get straight to the point so you can go home and get some rest."

"Thanks, Leslie. I look that bad, huh?" Leslie mouthed "sorry", then Susan continued. "I hate to add more to your plate, but Anthony at the module site needs help. He and I both tried to get help from Timothy, to no avail. The contractor is under schedule pressure and is not handling it well. They've had two serious near misses in the last week, and I can see more coming based on what I observed at the site."

"Oh, I see. First of all, thanks for going out to the site on short notice and for coming to me so quickly. When I reflect on the past couple of months, schedule has become a larger and larger part of my discussions with Timothy. That module is holding up the entire project. But I've been clear with the entire leadership team about our commitment to safety. Perhaps I sent conflicting messages."

"Anthony clearly received those earlier messages. He can't resolve the dilemma himself. I know from experience you won't hesitate to roll up your sleeves and help when needed."

Leslie said, "Let me talk to Timothy. We're considering even more work with that contractor. Perhaps we can use that as an incentive to improve their resourcing and safety behaviors. I'll let you know how it goes."

Leslie had made her safety expectations known throughout the project and had reinforced them with well-placed interventions. Team members thousands of miles away had noticed. But she let up with the increasing pressures to complete the project on time. She knew she was doing so, but assumed her credibility would carry them through the short remaining time. It did for most of her team, but Timothy interpreted it as support for delivering at any cost. It just took her HSE manager holding up a mirror to realize she had erred in judgment. She needed to roll up her sleeves again and rectify things, both with the contractor and Timothy. She would also reinforce the priority of safety with the rest of her leadership team.

But what if Leslie had ignored the feedback or denied a problem? She might get away with it if they finished the project on time without an injury. But if someone was seriously injured, her recent behaviors would likely be uncovered in a thorough investigation. Her credibility as a safety leader would be based on those last few months versus the years before. Susan would have been deflated. It took a lot of courage and energy to deliver the message, especially after being rebuffed by Timothy. Her sacrifice would have been for naught.

How well do you receive such feedback? Do you seek to understand, deny, argue, ignore? How do you encourage your team (and others) to provide open and honest feedback? Have you ever delivered such feedback to your boss? How did they respond?

***

In the following sections, I describe a variety of methods to obtain such feedback. The best ones for you will depend upon your credibility, how you have responded to feedback in the past, and the culture and size of your organization. They range from simple activities you can do yourself to formal, externally facilitated options.

Self reflection

While a leader can't see it all themself, they can find many unintended messages through deliberate self reflection. In fact, the self reflection questions in this book are intended to help leaders practice this skill. It takes time and discipline and works best for humble leaders. Leaders' agendas are often so full that they move on to the next engagement or activity without reflecting on the impact of what they just did.

This can be done before or after engagements, actions, or decisions. If done beforehand, a leader can anticipate potential misinterpretations and adjust their words or action ahead of time, thereby eliminating potential damage to their credibility. Unfortunately, they will miss a valuable input, the reactions of the people around them: the questions, the facial expressions, and the murmurs in the crowd. In addition, they won't be able to reflect on encounters that occur in the spur of the moment. These gut reactions show the true values of leaders, and people know it.

If done afterwards, the leader can analyze any of the engagements. They can consider the reactions of the audience. They can compare the words that came out of their mouth with what they intended. These are not always the same, just as I fail to capture all my intended messages with the words on these pages.

Reflection can be done in many ways. I will describe one systematic approach, but the consistency and quality of the reflections are most important here, not the specific method used.

Take ten minutes at the beginning of each day to reflect on some of your engagements or actions from the day before that had any potential for safety messaging. Focus mostly on the ones with the most at stake, but mix in a variety of others on different days to obtain a representative view of your messaging. Select some situations where other business drivers are the main subject of the engagement. This can be done at the end of the day as well, but I have more energy at the beginning of the day, and a night of sleep creates distance that increases objectivity. My best reflections of the day come outside of the work environment.

For each situation you select, reflect on a subset of the following questions. Feel free to add others that are more meaningful to you. This is a brainstorming exercise. Try to put yourself in the shoes of a diverse range of people you engaged with: direct reports, frontline workers, contractor representatives, those who adore you, and those who may suspect your intentions.

- What was your primary message during the encounter?
- What was the intended message on safety in the encounter?
- What safety messages could others have taken away from the situation?
- How did the audience or impacted persons react when you delivered the message or took the action? What do those reactions indicate about the message they received?
- What feedback did you receive from others following the encounter?
- How natural were your behaviors and mannerisms during the encounter?
- If you discovered any messaging inconsistent with your safety vision, what can you do to correct the misaligned interpretations?
- How can you improve your messaging in the future based on what you learned from this reflection?

Since this is only a reflection by yourself, you won't have all the answers. Note the items that concern you the most or you are most uncertain of. You can explore these further using some of the other feedback mechanisms below.

The questions above are written in the past tense for use after an encounter, but many can also be used ahead of planned encounters to help prevent unintended messaging. You can use some of the time at the beginning of the day to do this planning for some of the known encounters during the day.

Another way to facilitate self-reflection is through a more structured self-assessment tool, based on the list of behaviors that support or detract from your credibility as a safety leader, basically the material in Chapters 4 and 5. Provide examples of how you have demonstrated each trait. For example:

- What specific descriptions of your safety vision have you provided to the organization?
- What have you done to build your safety competence?
- How did you react following the last serious incident? How did you show care for the injured person? How involved were you in the incident classification?

You get the idea. You can obtain a consolidated list via the link in the Author's Note. You can reflect on one trait each week or go through them all every month or quarter, whatever works best for you.

Checking messaging with others

This approach can also be used before or after engagements or actions. Find a few trusted advisors you can use as a sounding board for your future or past encounters. You could use a person with a particularly diverse view or several with unique roles and perspectives. Potential candidates include safety leaders, your deputy, a representative of frontline workers, or a frontline supervisor. Make it clear that you are not only seeking their own view, but also how they believe others would react.

For planned engagements, meet with the person(s) ahead of time and deliver your planned message or share your proposed action with them, providing them only with the context that others receiving the message will have. Too much background information may prejudice their view. You can discuss the broader context after receiving their initial feedback. The questions are similar to those above, but adjusted for the different context:

- What was the primary message you took away from the words or proposed action?
- What do you think was my intended safety message?
- What safety messages did you take away from the words or action? How did you arrive at them?
- What potential safety messages might others take away from the words or action? Why do you think so?
- How consistent are the messages with your understanding or our safety vision or my safety commitment?
- What should be adjusted to reduce the likelihood of unintended messaging?

For analyzing engagements that have occurred, the same questions can be used. The last question can be replaced with:

- What can be done to correct the misaligned interpretations?
- What can I do in the future to prevent such unintended messaging?

Feedback circle

This option is similar to the one above, but in a group setting with a more open agenda. Instead of the leader driving the agenda, the feedback circle drives the agenda. This approach opens the possibility of receiving feedback on engagements where you did not expect mixed messaging and the cumulative impact of many smaller mixed messages. With this approach, you have the potential to receive multiple viewpoints in a short amount of time and people may feel more comfortable opening up with others present to support them. A disadvantage is that multiple people may latch on to one comment, giving it disproportionate attention and discouraging feedback on other issues.

Select a small (say 4-8) group of diverse representatives from the organization who have a reputation for providing open, constructive feedback. Meet with them regularly to build trust and rapport. Identify a few specific engagements since the last meeting to break the ice, if necessary, but give the group the space to provide the feedback on their mind. The leader should focus on what is on the group's mind, not their own. Some organizations may have existing HSE committees that can be used instead of forming a new group, if the size and makeup is right.

The questions from the last section can be used to probe specific engagements, but also have some open-ended questions in mind to solicit the feedback that can't be anticipated.

- What is the most powerful safety message (positive or negative) that you received from me in the past month? Why do you feel that way?
- What have I said or done that you or your colleagues view as inconsistent with our safety vision?
- How consistent are the safety messages you receive from me, my leadership team, and other leaders in the organization?
- What does our safety vision mean to you?
- What do you believe I expect you to do to contribute to our continued improvement?
- How can I inspire and empower people to improve safety performance?

As long as the leader is open to the feedback and acts on it, the very act of inviting the group in and listening builds credibility.

External coaching

Most of the options in this section use resources in the leader's organization. They know the work culture and business better, have seen their leader's behaviors over a longer time, and have a better context to provide feedback. Another option is to use an external (safety) leadership coach. This option may be best if a leader's credibility as a safety leader is low, in which case, they are unlikely to receive open or meaningful feedback internally. Many options exist, including combinations of the following:

- Regular phone calls with a coach to discuss recent engagements.
- Onsite shadowing to observe your behaviors, words, and actions and their impacts on others.
- Private interviews with a diverse group of people in the organization.
- An anonymous survey.

As is usually the case, the more a leader invests in one/one engagements onsite, the more valuable the feedback is likely to be. Remote coaches can help them through the reflection process, but don't know what they aren't told and cannot see the reactions of others. Feedback from surveys can be generic. More on that in the next section.

360-degree surveys

360 means receiving feedback from all directions: your boss, your direct reports, your peers, frontline supervisors, and even your business partners. The surveys can be done either internally or by external consultants. Not all external surveys are created equal, but in general, they are designed to produce reliable feedback and can be benchmarked with other leaders in the same industry or beyond. The advantages of internal surveys are lower cost and the ability to tailor the language and questions to the business. With the use of online survey tools, either can provide anonymity, though some may trust a survey administered by a third party more.

I recommend at least two components of a survey, specific questions aimed at traits and actions (such as the ones described in this book) and open-ended questions for survey participants to describe specific examples. Most participants, especially those who are weary of surveys, will only answer the specific questions, so a leader should encourage participants to provide more information in order to learn more effectively. The first part of the survey will typically only identify general strengths and weaknesses. The leader may struggle to reconcile the feedback without the specific examples given in response to the open-ended questions. If nothing else, these surveys can give the leader a focus area or two to direct their other feedback mechanisms. The surveys may also provide a semi-quantitative measurement of their safety leadership over time.

Following such surveys, it is important for a leader to recognize the contributors and reinforce their commitment to learn and improve. They may not be able to review the feedback with everyone, but should consider doing so with their leadership team, who may help fill in some of the blanks the numbers don't tell.

Informal feedback

Obtaining feedback doesn't always have to be so formal. The techniques listed above are more likely to provide a leader with more complete and reliable feedback. However, leaders can use their hundreds of encounters each week to gather feedback as well. The requests can be ad hoc in random encounters. During performance reviews, they can ask for feedback on their own performance. Many leaders can use their leadership teams as sounding boards. The opportunities are only bound by the hours in the day–and the patience of those who are asked to help.

Act on the feedback

All the efforts to collect this feedback will be for naught if a leader doesn't act on it. In fact, if their credibility is already low, inaction may cause their credibility to fall further. Their humility was not sincere; they refused to learn; they didn't know how to improve, etc. By act, I don't mean a leader should carry out every suggestion provided to them. Acting can be providing feedback on why they are not taking action on the suggestion or perhaps taking an alternative action. Take actions that matter, not just to improve credibility. Otherwise, they may be seen as a tick the box exercise for the leader's own benefit.

An organized leader committed to improvement will record feedback received and actions to be taken (or not, with justification) in an action log or in their individual goals. This can be reviewed with their feedback circle, if they have one, to confirm they are working on the right things and the actions are having the desired impact.

In most cases, the changes in a leader's behaviors and actions should speak for themselves. Reporting out on them to the broader organization may be viewed as self-centered. However, if credibility is very low and the improvement process is formal, having these open discussions may have no downside and may show humility and a genuine desire to improve.

Summary

- Poor leaders are often blind to the impacts of their words and actions on others
- Seek feedback from others using one or more of the methods described here.
- The extent and quality of feedback needed will depend on your level of credibility and your appetite for learning and improving.
- Act on what you learn.

Personal reflection and learning

- How much do people in the organization trust you? How much does it vary across the organization? How will it enable or limit the use of these feedback mechanisms?
- Think of occasions where you received feedback on your safety leadership behaviors. How was it solicited? How did you react?

- Name six people you trust to provide open and constructive feedback on your safety leadership behaviors. Why did you select them? Would they provide better feedback individually or as a group?
- Practice a self-reflection exercise as described in the first section above.
- Conduct a more complete self-assessment using the worksheets referenced in the Author's Note

9

Regaining Lost Credibility

Credibility is hard to earn, easy to lose, and even harder to get back. A leader can lose it by taking a significant action inconsistent with their professed safety vision or through a series of less significant actions not aligned with the vision. The latter is less likely because the propensity for such behaviors means the leader doesn't have the consistency needed to build credibility in the first place. I will focus on the former here.

How can a leader regain the credibility they once had? Let's be honest, it is difficult, and sometimes impossible. As I said earlier, people expect a leader to get the easy engagements or decisions right. The true tests are the dilemmas where two or more of their top values collide. Recovery is easier for those with extensive demonstration of sincerity and consistency. Those who have shown some evidence of doubt in the past face a more difficult recovery.

The first step is to review your HSE vision and targets. If you no longer believe in them personally, or never did, you probably won't recover. Consider revising them, or better yet, working with your organization to do so. The disconnect is probably what caused you to lose your credibility in the first place and will prevent you from taking the bold and consistent steps to recover it. You will have little room for substantial revision with lives and livelihoods at stake. It still needs to be inspiring and worthwhile, and one where your gut reaction is the right one.

Once you revise your expectations, recommit, and share with your organization, you can move on with recovering your credibility. So, how do you do that? There is no recipe or standard operating procedure. As a follower, I would look for the following:

- Admit your mistake(s) and show intense humility.
- Take a bold action (or two or three) consistent with your safety vision.
- Seek help from others and act on it.

These are only the beginning. People will expect you to demonstrate behaviors consistent with the vision for a long time before restoring their trust. All will not be forgiven by a genuine apology or a single promise. Fortunately, these three recovery measures fit together well. They have been covered in detail in previous chapters, so I won't dwell on the basics here. It makes sense that the traits and actions which make many leaders feel the most vulnerable are the ones that can help them recover credibility.

Show humility

Admit your mistake. Acknowledge what you did was inconsistent with your safety vision. Explain why you did it. Acknowledge the actual or potential consequences. Whatever you do, don't make excuses; that will sabotage your recovery efforts immediately.

Act boldly

Can you undo the wrong deed? You can't take back insensitive words, but you might be able to reverse a decision. Can you at least prevent some of the negative outcomes of your action or decision? If not, seek another opportunity to take bold action to recommit to the safety vision. The sooner, the better. Look for an opportunity versus waiting for it to come to you. All the words in the world will seem cheap and artificial until bold action backs them up.

Seek help

I covered this in the last chapter. In a recovery situation, the more visible approaches will work best, i.e., an external survey plus coaching and/or a feedback circle. In this case, it is useful to share summaries of the feedback you are receiving and your specific actions to address the feedback.

***

Let's return to the story about the decision to leave the crew on the offshore drilling rig with an approaching hurricane.

> Carlos, the rig manager, stepped off the helicopter and braced himself for a chilly reception. He had a few minutes to get settled and talk with Terry before meeting with the crew during lunch. Terry had told him to prepare for the worst.
>
> The marine supervisor walked in. "Oh. Hi Carlos. I didn't know you were here already. I'm looking forward to our meeting. I don't know what you can say to make up for leaving us out here during the storm. My wife was scared to death as the storm came onshore. It was her first time alone during a hurricane. I felt so helpless out here."
>
> "I'm so sorry, Perry. I messed up. I let you all down when I was the only one that could help. Unfortunately, I can't undo that, but I can apologize and do better next time."

"Yeah, but what are you going to do? I hear the ERT makes those calls."

"For one, I joined the ERT, so I will not only be a part of the decision for the entire region, but I will seek input from the other rig managers. And second, I just told Terry that I'm setting up a safety feedback circle. I do this informally already, but I want to give the crew a more formal opportunity to provide feedback directly to me on improving safety out here, not just for emergency preparedness. You know I've been a big supporter of safety in the past."

"That's a good start. Sign me up for the group. It will take time for you to earn back our trust, but I have to admit it's safer out here now than five years ago." Perry turned to walk out of the office.

Carlos grabbed Perry's arm. "By the way, here's a gift card for you to take your wife out for a nice dinner during your days off. I won't get to see her in person, so this is at least a token of my apology to her as well."

Summary

- Credibility is hard to earn, easy to lose, and even harder to earn back.
- You may be able to earn it back through humility, bold action, and seeking help.

Personal reflection and learning

- Describe a situation where you saw a leader (or yourself) lose their credibility as a safety leader. What did they do or say? How did they respond? How did they try to recover? If unsuccessful, what else could they have done?

- Review your safety vision. Are you still aligned with it? Really?

10

Leadership Alignment

The focus of this book has been to help a leader of a large organization. In reality, safety excellence is not so simple. To achieve the almost perfect performance many desire, all leaders in an organization must be aligned with the safety vision, with their words and actions supporting each other.

A few stories in this book have shown how misalignment of leaders at various levels or in different parts of the organization leads to skepticism and frustration in the workforce. Recall the story about Timothy, Anthony, Susan, and the process module being built overseas. Timothy and Leslie weren't aligned on the safety vision, which led to Timothy going astray on safety when schedule pressure rose. In the story about the preparation for a hurricane on the drilling rig, the onshore ERT let the people offshore down, despite the monumental efforts by Carlos, Terry, and others to build a strong safety culture.

Leaders are stronger in demonstrating some traits and weaker in others. They have their own styles. For the most part, this is good. This personalization demonstrates a degree of sincerity. They are not following a recipe given to them by others, nor are they ticking boxes. They are being themselves. Forcing consistent styles would make them appear unnatural and contrived, perhaps signaling a lack of sincerity. However, individuality can go too far when it prevents leaders from adopting better practices from others.

The messages received from all the leaders must be consistent with the organization's safety vision. If not, people will become confused and follow their natural instincts, whether consistent with the vision or not.

In this chapter, I list actions that leaders in various roles can take to build this consistency. This is not only for leaders as designated by an organization chart; all members of the organization are leaders in this context. I will not repeat all the actions covered earlier; rather, I will list some actions that may vary by role. I hope you will see how they can complement each other.

Top leader of an organization

- You should know this one by now. See Chapters 4 and 5 for dozens of actions.

Middle management

- Provide input on the safety vision when requested (or not). Represent the views of your part of the organization, not just your own.
- Tailor the implementation of the safety vision to the work done by your part of the organization.
- Develop an action plan for your part of the organization to contribute to annual safety goals and targets.
- Spend time observing and discussing the practical application of the vision in people's everyday work.
- Help frontline supervisors resolve dilemmas; share samples with your boss.
- Seek feedback on your own safety messaging.
- Provide feedback on safety messaging of others, including your direct reports, peers, and boss.

Frontline supervisors

- Provide input on the safety vision when requested (or not). Focus your input on the impacts on the frontline workers.
- Tailor the implementation of the safety vision and your department's action plan to the work done at the frontline.
- Verify implementation of hazard controls at the frontline.
- Help frontline workers resolve dilemmas.
- Seek feedback on your own safety messaging.
- Provide upward feedback on safety messaging of others.
- Collect feedback of frontline workers and share with your middle management, regardless of how well it aligns with your own views.

Frontline workers

- Seek to understand the safety vision in detail. Ask if you don't understand it or how it applies to your work.
- Discuss dilemmas between safety and other priorities with your supervisor.
- Provide input to a member of the HSE committee, feedback circle, or similar group, especially if uncomfortable providing such feedback to your supervisor.
- Stop work when you observe behaviors or conditions inconsistent with the safety vision and discuss with your supervisor and peers.
- Provide honest feedback when approached by leaders visiting the worksite. Show your work, challenges, dilemmas, and pride.
- Share your ideas for improvement with your supervisor and safety staff.

Individual contributors

- Seek to understand the safety vision in detail. Ask if you don't understand it or how it applies to your work.
- Discuss dilemmas between safety and other priorities with your super-

visor.

- Provide input to a member of the HSE committee, feedback circle, or similar group, especially if uncomfortable providing such to your supervisor.
- Seek input from frontline supervisors when planning work or developing or modifying systems they use.
- Provide upward feedback on safety messaging of others.

Safety department members

- Provide options to the line leaders developing safety targets and improvement plans.
- Insist that line leaders own safety.
- Seek feedback on your own safety messaging.
- Deliberately observe the potential and actual impacts of words and actions of leaders. Provide constructive feedback. Escalate when you cannot resolve yourself.
- Execute or manage feedback surveys.
- Adjust safety-related systems, procedures, and processes to align with the safety vision and to address feedback from the frontline and others.
- Provide safety performance summaries to line leaders, including leading and lagging indicators. Ensure the data is accurate and representative of the work.

Thought leaders

- <u>Note</u>: Thought leaders have no formal authority, but have an extraordinary passion for safety and strong influence over others in the organization.
- Engage with line leaders to deepen your understanding of the safety vision.
- Advocate the safety vision within your circle of influence.
- Deliberately observe the potential and actual impacts of words and

actions of leaders. Provide constructive feedback.

- Contribute to improvement efforts outside of your core responsibilities.
- Seek feedback on your own safety messaging.

Business partner leaders

- Seek to understand the Company's safety vision at the appropriate level. Compare it to your own. Communicate and resolve differences with the Company.
- Engage with your own leaders and workers to help them understand the interface between the safety visions and programs of your company and the client or partner. Keep it simple.
- Seek feedback on your own safety messaging, both from your employees and the client/partner.
- Provide feedback on safety messaging of others, both in your company and with the client/partner.
- Revisit alignment in contract review meetings with the client.

Summary

- All leaders in the organization must act in alignment with the safety vision to achieve excellent safety performance.
- Provide input to the development of the safety vision and targets, whether asked to or not.
- Provide feedback on observed safety messaging of leaders: upward, downward, and sideways.
- Share dilemmas between safety and other priorities. This is a sign of strength, not weakness.

Personal reflection and learning

- What examples of inconsistent safety messaging have you seen in your organization? What effect did it have (or could have had)? What did you do about it?

11

Conclusion

Sending people home every day without harm is the most difficult aspect of most leaders' jobs, especially those whose organization is large and executes high risk work. Harder than growing the business, harder than reducing cost, and harder than finishing a project on time. Unlike other aspects of the business, failures in safety can cause irreparable harm. Lost market share can be regained. Losing a life is forever.

This book was targeted at leaders who aim for excellent safety performance. Only they and their organization can define what that means specifically, but they must do so in order to guide their words, actions, and behaviors. It could mean anything from no paper cuts to no fatalities. If it is not clear in their own mind, their behaviors are likely to send mixed messages, resulting in mixed performance and injuries. In order to achieve safety excellence, everyone needs to be aligned with the vision and understand how they can and should contribute.

Defining the safety vision and expectations is only the beginning. A leader must consistently demonstrate they are sincere in their mission. Only then will they be a credible safety leader. *Sincerity + Consistency = Credibility.* Credibility comes with trust. People are more likely to follow their leader on the difficult journey ahead. Those without credibility face apathy, skepticism, or distrust and find making positive changes difficult, if not impossible.

Leaders are under the microscope of people in their organization every day. People are looking for messages behind each sentence and decision. The traits covered in this book are much of what people are looking for. I described practical actions and behaviors to help you demonstrate those traits. I used stories to illustrate many of the traits and actions. I held a mirror to your face. If you were not able to open your mind and reflect on your own behaviors, I encourage you to reread the stories now that you have seen the full context of the book.

If you are still unsure of what to do next, you can use the self reflection worksheets referenced in the Author's Note to identify your strongest and weakest traits. You don't have to be excellent in all, but strong enough to avoid contradictory messages. A self assessment only takes you so far; set up an ongoing feedback mechanism to continue learning and improving.

Know the Talk! Walk the Talk! Listen to the Talk!

Best wishes on your journey. There couldn't be more at stake. The hard work will all be worthwhile.

Author's Note

I hope you found this book enjoyable, motivating, and useful in planning your next steps to become a better safety leader. Please **leave a review** where you bought this book to help others discover the book and become better safety leaders themselves. Thank you in advance.

You can download your bonus Self Reflection Worksheets referenced in the book here. (https://drive.google.com/uc?export=download&id=1_Kzsw0N NeluWGEs4YLgGedoCP7Ocqw8j).

For updates on new releases, free bonus material, and special deals, please sign up for my newsletter here (https://strivingforsafety.mailerpage.com). You can unsubscribe at any time if you don't find them worthwhile. Please encourage others in your network to read the book and sign up for the email list. This leads to more excellent safety leaders, from which we can all learn from later.

My website is https://strivingforsafety.mailerpage.com/ and my LinkedIn profile can be found here (https://www.linkedin.com/in/arnold-marsden-1 4170180).

You can email me at arnold@strivingforsafety.com. I love to receive feedback from readers on my books and their best practices.

Best wishes on your journey!

About the Author

Arnold lives in Fulshear, Texas, USA, just outside of Houston. He worked as a Health, Safety, and Environment (HSE) professional for over thirty-five years helping to prevent injuries, save lives, and protect the environment. Since retiring, he continues to hike and backpack in some of the most beautiful parks and wilderness areas in the country and has begun an author career, writing both fiction and nonfiction.

In his nonfiction books on industrial safety management, the Safety Through Story series, he shares valuable lessons learned from his career to help others save lives. His first book was *Don't Let It Fall: Stop Dropped Objects, Save Lives*.

Muir Trail Magic is his first novel in the *Bucket List Hike* series and was inspired by his own hike of the John Muir Trail in 2021 and 2022. Stay tuned for future releases following Bob, Cathy, and others on hiking, backpacking, and sightseeing adventures in the National Parks and other nature preserves. Though fictional, he hopes that the books help you prepare for your own adventure, relive a past adventure, or experience the area from the comfort of your own home.

To learn more about his books and subscribe to his author newsletter, visit his website. (https://strivingforsafety.mailerpage.com)

Acknowledgments

Thanks to all my former colleagues and business partners for sharing and listening.

Thanks to those working at the front-line, producing the goods and services we all need while protecting themselves from an endless array of hazards.

And a special thanks to Rick Strouse and Eden Newell for reading a draft and providing valuable feedback to make it a better book.